# The Living Barrier

# The Living Barrier

## A primer on
## transfer across biological membranes

**Roy J. Levin, B.Sc., M.Sc., Ph.D**
*Senior Lecturer, Department of Physiology,*
*University of Sheffield*

**William Heinemann Medical Books Ltd**
*London*

*First published 1969*

© Roy J. Levin, 1969

SBN 433 19240 2

*Printed in Great Britain by*
*Cox & Wyman Ltd, London, Fakenham and Reading*

If not for
Anne, Chaim and Margaret,
then for who else?

# ACKNOWLEDGEMENTS

Hidden behind the print of most books are the helping hands and thoughts of many people. This book is no exception to the rule. It would have been much more difficult without the aid of a number of people whom I would like to acknowledge and thank.

First comes Miss Dawn Hinde who typed and retyped the manuscript without any grumbles over my many changes of wording and content. Second comes a group of colleagues of the Department of Physiology at Sheffield University. They read the manuscript and offered many helpful suggestions and comments. Top of the list is Dr Peter Kohn, then comes Frances Edwards, Derek Hudson and Kevin Barkla. Last, although by no means the least, comes Miss Joy Anderson who drew the diagrams from my rough sketches.

Dr Ian Carr, of the Department of Human Biology and Anatomy at Sheffield University and the Royal Microscopical Society, generously gave me the photograph of blood cells taken by the scanning electron microscope.

# LAST WORDS UNDER THE GUISE OF A FOREWORD

I once read that a foreword is always the first thing in any book that is read last. Furthermore, though they constitute the beginning of books they are always written last. Bearing these facts in mind forewords are clearly misnamed and should be more accurately titled "last words". In truth they are often an author's comments after he has finished his manuscript and thus become an apology for some of the "warts and pimples" that he now sees have crept into what was initially going to be a flawless blend of mind and pen.

In a small book about a subject as large as biological membranes some guiding ideas about the length (apart from the publishers economic ones!) and the level are needed. So in order to absolve myself from irate readers who found that I had left out this aspect of membranes or that "there was only one sentence on that aspect" I have generally tried to write the book in the light of the teaching rules of the great Czech educator Comenius. They are – "The few before the many, the simple before the complex, the general before the particular, the nearer before the remote, the regular before the irregular, the analogous before the anomalous." If the book does not always come up to these stringent statements the fault is mine not Comenius's! Some readers may be surprised that this primer is written where appropriate in a semi-narrative style, as if the study of membranes was a continuous story of human activity. The choice was deliberate. Our ideas on cell membranes are constantly changing. They come from good and bad experiments; from guesses, hunches and speculations; from accurate and sometimes inaccurate measurements; from purely theoretical calculations and from purely descriptive studies. Each fact that we now accept as "hard" information, as practically a self-evident truth, has come about from the endeavour of human curiosity channelled into scientific study and research.

Although the book is intended to be a primer I have not hesitated to include ideas and experiments that may seem to some people too advanced for the readership. It could be argued that I am going against Comenius with his "simple before the complex". I do not think, however, that because an experiment is a little complicated and needs careful thought that it must be automatically excluded from an elementary text. On the contrary, it has been my experience to find only too often that elementary texts, by scrupulous avoidance of anything complex and advanced, grossly oversimplify and often become inaccurate and long-winded. Furthermore, I do not believe that because a book is a primer, that all vestiges of controversy, criticism and doubt about the subject matter must be removed. By making an introductory text full of "cast-iron facts" it gives a false impression that all is clear cut and known. The next stage in learning then becomes one of "unlearning" the previous so-called cast-iron facts! Such an approach must stifle that precious commodity, scientific "creativity". A primer should show paths to follow, with reasons for choosing a particular one, not simply lead the reader blindly down the straightest.

To state this in a foreword is clearly dangerous for it invites the reader to be even more critical and watchful of the text that follows. If, however, the initial argument that forewords are read last is correct, this may not be too worrying a detail! All I can hope is that I have avoided at least some of the pitfalls I described.

*November 1968*                                Roy J. Levin.

*Department of Physiology,*
*University of Sheffield*

## CONTENTS

# CHAPTER 1

# The Discovery of the Cell

**Why study membranes?**

Like many other things in life the scientific investigation of biological processes has its fashions, for a variety of reasons some areas of study become more popular than others. At the present time molecular biology is the queen of the life sciences. Enormous progress has been made in the last ten years in our knowledge of the helical nucleic acids DNA and RNA and how these macromolecules carry the genetic code and control the synthesis of specific proteins. This has led to our understanding the basic mechanisms of cell duplication. Although such matters are the present focus of attention in the biological world the study of cell membranes runs a very close second. In fact, as the areas of scientific research are never static we may well see the study of membranes overtake the position of nucleic acids in the "number one spot".

The way membranes control and influence the movement of substances into and from cells are topics that fill the waking hours of many biochemists, physiologists, pharmacologists and botanists. The solution of such fundamental biological problems holds the key, not only to the understanding of fundamental processes of life but to many spectacular advances in medicine and other biological sciences. The outer cell membrane is the first contact of life with its environment and is thus the first barrier between life and changes in the external environment. Because of this simple fact the properties of this living barrier must be of supreme importance.

**A profusion of membranes – part one**

When we examine the term membrane we find, unfortunately, that it has been used for more than one type of biological structure. Currently it is applied to three different types of systems (Fig. 1).

1. The term is used for the external membrane of an individual cell. This external membrane is often called the plasma membrane or cytoplasmic membrane or plasmalemma.

2. Single layers of similar cells lining an organ are often called membranes. For example those lining a capillary are sometimes referred to as the capillary membrane.

3. Complete epithelial structures involving many different cell

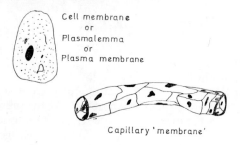

Cell membrane
or
Plasmalemma
or
Plasma membrane

Capillary 'membrane'

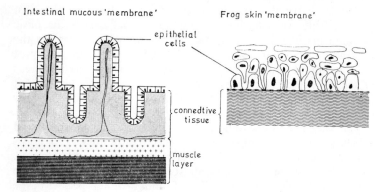

Intestinal mucous 'membrane'                    Frog skin 'membrane'

epithelial
cells

connedtive
tissue

muscle
layer

*Fig.* 1. The various biological structures to which the term "membrane" can be applied. The term is used for the outer membrane of the cell, for single layered sheets of like cells (such as those that line and form the capillary wall) or for epithelia that contain many types of cells, e.g. the intestinal mucous membrane and the frog-skin membrane.

types but which basically act as a barrier between compartments are occasionally called "membranes" for convenience. Thus one talks about the intestinal "mucous membrane" or the gall-bladder "membrane" or the frog-skin "membrane". It is often assumed that the properties of these epithelia are largely controlled by the properties of the individual plasma or cytoplasmic membranes constituting the cells. Similarly, because isolated cells are not easily prepared, whole epithelial tissues containing large amounts of the plasma membrane under study are often used for experimental purposes. The data obtained with these "membranes" are thought to reflect the properties of the individual plasma membranes. Sometimes, however, this is not so, the whole being more than just the sum of its parts. Structures in biology arising from a number of isolated units often have properties that the single units do not possess. Examples of this for membrane systems will be dealt with later in the text.

### The function of cell membranes

Structure is always related to function. This dogma of biology holds not only for the gross structures like bones, organs and tissues but even for cells and their subcellular components. Membranes undertake many different functions despite their apparent simplicity of structure. The diversity of function is worth listing.

1. **Supportive functions.** The cytoplasm of most cells is an aqueous phase with little form and much plasticity. The many membranes of the cell, both internal and external, maintain its structural integrity.

2. **Segregative or compartmental functions.** Membranes can either keep substances close together or can keep them apart. Association or dissociation of enzymes from their substrates or products is clearly important in many cases. Chains of enzymes have to work in an integrated manner with one another when association is of the highest importance. The best example of this is the enzyme system in mitochondria which generates ATP and carbon dioxide from glucose and oxygen. Molecular and structural co-ordination is exemplified in this organelle where substrates and electrons have to be rushed along chains of enzymes. The importance of dissociation of enzymes from substrates is best shown by the packaging of powerful hydrolytic enzymes into small membrane bound bags called lysosomes. These enzymes are so powerful that they would tear the fabric of the cell apart if released inadvertently (hence their nickname of suicide bags!). They are compartmentalized into vesicles to prevent them from attacking the substrates present in the cytoplasm.

3. **Exchange functions.** Membranes act as barriers to regulate the movement or exchange of substances between compartments. Because membranes restrict free diffusion, special mechanisms

have often to be provided to allow nutrient substances through selectively.

4. **Catalytic functions.** Many membranes have been found to contain catalytic activity usually associated with enzymes, while some enzymes are known to be bound into the membrane structure. The microvilli of the brush border of the mucosal cells of the small intestine contain enzymes that break down disaccharides and peptides into their constituent monosaccharides and amino acids respectively. These disaccharidases and peptidases can be removed from brush borders by treatment with the proteolytic enzyme papain, suggesting that they are intimately bound up with the structural proteins of the membrane of the microvilli. It is not only in animal cells that enzymes are involved in membrane structure. Bacterial membranes contain enzymes (permeases) that play a major role in transferring substances into the bacterial cell. The chloroplasts of plants synthesize sugars from carbon dioxide and water using the energy of sunlight and the catalytic powers of chlorophyll. The metabolic enzymes for the synthesis are bound up with the chloroplast membranes.

5. **Movement.** Numerous types of wandering cells (unicellular organisms, leucocytes) move by complex interactions between the cell cytoplasm and the membrane structure. Even cells that are usually accepted as being static, because they are fixed in tissues, often show shimmering movements and undulations of their membranes.

6. **Attachment or adhesion.** How cells stick together to form tissues and organs is extremely complex. It is known that special modifications of the surface allow some types of cells to attach themselves to one another with considerable force. Kidney tubule cells and mucosal cells of the small intestine show modifications of their plasma membranes called "desmosomes". The cytoplasm appears dense around these areas (Fig. 13) and such strong bonds are formed that if the cells are pulled apart the desmosomes of individual cells stay attached to one another even though the cells may rupture. In other cells that do not have these structures it is thought that intercellular cement or adhesive is secreted on to the surface and that the calcium ion plays an important role in the adhesive action. Because proteolytic enzymes (like trypsin) are often used to separate tissues into individual cells, it is thought that this cement is proteinaceous.

7. **Recognition.** For a complex animal organism to remain com-

posed of its own cell type and to keep the cell population pure, the cells present must be able to recognize each other. Leucocytes have to be able to differentiate between other blood cells and bacteria otherwise they would engulf all cells without discrimination. The cells of the reticuloendothelial system in the liver (Kupffer or Stellate cells) or in the spleen must be able to recognize old worn-out red cells so that they can be destroyed. Here one set of cells is discriminating between young and old cells of another line, proof indeed of the exquisite sensitivity of the recognition mechanism. As yet we have little idea how such mechanisms operate but it is thought that they are dependent on chemical groups on the outer parts of the cell. These groups might also be the structural entities that give rise to antigen–antibody reactions and be the trigger for the immune response. The science of immunology has rapidly grown to accommodate such studies. Its impact on our society is seen in the headlines every time an organ like the heart or kidney is transplanted from one human to another. Whether or not the old tissues will reject the new cells of the transplanted organ often becomes, for short periods of time, the concern of whole nations.

### The evolution of the cell concept

To begin at the beginning is perhaps the most obvious way of approaching the study of the cell membrane. To do this, however, we must start with another history, that of the evolution of the concept of the cell as the unit of life in animals and plants. It is difficult today to appreciate the slow struggle that was needed to develop this concept which is now taken for granted. No single person can be credited with this fundamental discovery. Its long history depended not only on advances in the techniques of handling and looking at biological material but also on advances in physics and chemistry. Changes in the religious thoughts and attitudes about man's study of his body and its processes also played no small part in the evolution and development of the cell concept, some landmarks of which are summarized in Table 1.

### Early observations of biological material

The development of optical methods to magnify small objects have their origins in antiquity. Euclid (590 B.C.) investigated the properties of curved reflecting surfaces. Seneca (A.D. 65) used glass spheres filled with water to magnify objects while Ptolemy (A.D. 127)

TABLE 1. *Some landmarks in the development of the cell theory*

| | |
|---|---|
| A.D. 65 | Seneca found that water-filled glass spheres magnified objects. |
| 1235 | Bacon constructed magnifying glasses and manufactured spectacles. |
| 1485 | Leonardo da Vinci advocated lenses for looking at small objects. |
| 1590 | Jans and Zacharias Janssen invented the compound microscope. |
| 1665 | Hooke described cork structure and used the term "cells". |
| 1674 | Leeuwenhoek drew the protozoa, bacteria, rotifers and red blood cells observed under his simple lenses. |
| 1780 | Adams and others devised machines that sliced tissues very thinly (microtomes). |
| 1809 | Lamarck thought that cellular tissue was the general matrix of plants and animals. |
| 1824 | Dutrochet argued that tissues were globular cells held together by adhesive forces. |
| 1826 | Turpin proposed the cell as the fundamental element in the structure of living bodies. |
| 1831 | Brown described the nucleus. |
| 1838–39 | Schleiden and Schwann asserted that "elemental parts of tissues are cells similar in general but diverse in form and function . . . the essential thing in life is the formation of cells". |
| 1858 | Virchow realized that diseases were due to affected organs, tissues or cells and that "Every cell must come from a cell" (Omnis cellula e cellula). |
| 1861 | Schultze established the protoplasm concept by stating "the cell is an accumulation of living substance or protoplasm definitely delimited in space and possessing a cell membrane and nucleus". |

studied the magnifying effect caused by curved surfaces. Although such pioneering observations helped to lay the foundations of optics for subsequent developments, real advances did not begin until much later. Roger Bacon in 1235 constructed magnifying glasses and manufactured spectacles while Leonardo da Vinci, always far-sighted, suggested in 1485, that lenses should be used to study small objects. By the time the middle and late seventeenth century had arrived simple hand lenses had become an amusing toy for the rich and well educated to play with. We know that Galileo used one to examine ants, moths and fleas and sent one to a friend with a note "Herewith an occhialino for examining minute things at a near distance. I hope you will have as much use and joy as I . . .". The lenses created enormous interest but they had a number of serious disadvantages. They were usually little more than small globules of fused glass or crystal with a very small field of vision and must have caused great eye strain when in use. Because their surfaces were irregular the images formed would be distorted and have the coloured rings or halos around them caused by chromatic aberration. It is no surprise then that the flurry of excitement and amusement caused by their use died away. Previously two Dutch brothers, Jans and Zacharias Janssen, realizing the limitations of single lenses combined two in a tube to make the first crude, complex microscope in 1590. Their invention was ignored and many years passed before its advantages were realized. Remarkably, the great epoch making observations of Robert Hooke in England and von Leeuwenhoek in Holland were all made with the simple, tiny single lenses. The young Hooke, after a mere two years of biological studies, published a book about his experiments in 1665. The ones that have left a lasting mark in biology were his microscopical observations and his concepts on the structure of cork. In the book called *Micrographia*, he used the term "cell" to describe the little boxes lined by the thin dead walls of cork. The word was later modified by others to mean the living unit of tissues rather than the dead walls of the cork bark. Some twelve years after Hooke's publication Anthony von Leeuwenhoek, a linen-draper in Delft who had become an ardent amateur biologist, published a series of drawings from observations with his lenses. What made Leeuwenhoek's observations outstanding was that his lenses were magnificent. Although he used only single, convex lenses he became an expert at grinding their surfaces, making them optically superior to any others. His enormous skill was matched only by his curiosity for he examined practically anything that he could lay his hands on. Fish blood, saliva, seminal fluid, pond and rain water all came under magnified scrutiny. His drawings of saliva indicate that he clearly saw moving bacteria. It would be nearly another 100 years before anyone else would study them – a clear credit to his powers of observation and his lens-grinding skill. He described and drew protozoa, rotifers, red blood cells and the human spermatozoa. He confirmed Marcello Malpighi's discovery of capillaries (1661). Some of his drawings of animal red blood cells show that he was the first to observe the intracellular body we now call the nucleus, but it was likely that his lenses, although able to give a magnification of up to 200 times, did not allow him to view it clearly nor to appreciate its structure. Leeuwenhoek guarded his best magnifying glasses from prying eyes and when he died his secret of grinding fine lenses died as well. With his death significant advances in the microscopic study of cells ceased. The trail was not picked up again until the beginning of the nineteenth century. Technical skills had by then reached a stage when machines were being devised to slice tissues thinly, a prerequisite for accurate observations of tissue structure.

Published descriptions of animal and plant morphology gradually began to consider that the large organs might be made up of smaller units. Jean Lamarck thought that cellular tissue was "the general matrix of plants and animals". Henri Dutrochet (1824) extended this idea a little further by arguing that plant and animal tissues were globular cells held together by adhesive forces while Turpin proposed that the cell was a fundamental element in the structure of living bodies as early as 1826. In spite of these attempts to establish a cell theory the scientific world took little immediate notice. The stage was set, however, for a definitive statement about the cell and its place in animal and plant anatomy. This came about by the work and meeting of two Prussian scientists, Matthias Schleiden, a former lawyer who turned to study the anatomy and physiology of plants and Theodor Schwann, an anatomist who specialized in zoology. Schleiden had published the idea that plants were made up of cells but erroneously thought that these cells came about by crystalization around the recently described nucleus that Robert Brown had discovered in plant cells in 1831. Schleiden made a visit to Schwann in the October of 1838. After discussing the cell theory in relation to plants, both scientists went to Schwann's laboratory where the similarity between animal tissues (actually those of a tadpole) and plant tissues was observed. Agreement was reached about the similarity of cell structure between the two great classes of living things. Schwann published a year later (1839) his historic assertion that "the elemental parts of tissues are cells, similar in general but diverse in form and function. It can be taken for granted that the cell is the universal mainspring of development and is present in every type of organism. The essential thing in life is the formation of cells." Considering the time taken for its development and the previous lack of interest, the theory rapidly became accepted by most scientists. Later, in 1861, Maximilian Schultze showed that plant cells and those of the lower forms were essentially the same, and that the basic unit or organization in every living being was a mass of protoplasm surrounding a nucleus. The ground substance of the cell was named "protoplasm" by Johannes von Purkinje, a Bohemian physiologist, about the same year as Schwann published his cell theory. Around the same time the great pathologist Rudolph Virchow published his Cellular Pathology. He wrote "that cellular theory applied to all living bodies leads us to cellular physiology and cellular pathology each based upon histology . . . the seat of illness must be investigated in the cell . . .". The cell theory had come of age. Its application to the phenomena of disease grew and it became the foundation of the new medicine that was to sweep Europe.

# CHAPTER 2

# The Cell Membrane Concept

## Cellular structure – early observations

Once the idea that the tissues and organs of animals and plants were made up of large numbers of individual cells was accepted, numerous observations with achromatic lenses on intracellular structures and their relation to cellular functions began. Attention was focused on the preparation of plant and animal tissues to preserve their structure as intact and as natural as possible for microscopic observation, i.e. fixation. The structures were then revealed by staining with various dyes and chemicals. The work was always of an empirical nature for the underlying chemistry of fixing and staining reactions was often unknown. To be fair much of it is still unknown and empirical even today. Such techniques, however, enabled many additions to be made to the simple statement of Schultze (1861) that "the cell is an accumulation of living substances or protoplasm definitely delimited in space and possessing a cell membrane and nucleus". Refinements in microscopy and staining allowed the once empty living substance or protoplasm to be filled with centrosomes (Theodor Boveri 1888), mitochondria (Benda 1897), the golgi apparatus (Camillo Golgi 1898), and numerous vacuoles, granules and vesicles filled with secretion. Even so, for the next fifty-odd years the pictures of the "general cell" published in textbooks were deceptively simple (Fig. 2). As late as 1949 a well-known student textbook of histology could state "that the cytoplasm is composed of an apparently undifferentiated part which contains the main mass of most cells". However, even the most modern advances in the design of microscopes with optical principles that allowed visualization of living structures inside the unstained normal cell (phase contrast and interference microscopy) could not add to the knowledge of the membrane structure, for even with perfect lenses the microscopist could never hope to see cellular features separated by less than half a wavelength of light (about 2500 Å).

The early history of biological membranes rested entirely on functional studies of cells. The concept of a cell membrane was continually inferred from experimental data without any anatomical proof.

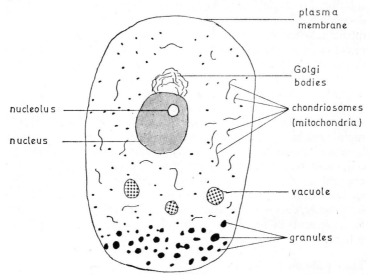

*Fig.* 2. The constituents of a hypothetical general cell as envisaged in 1925. The diagram is a schematic version of one appearing in a standard textbook of that year. Compare with Fig. 13.

### Early functional studies of cellular plasma membranes

A number of studies in the mid-nineteenth century indicated that animal electricity was generated between the inside and the outside of cells. Only ten years after the publication of Schleiden and Schultze's cell concept, Emil du Bois-Reymond (a German physiologist despite the French name) noted that a voltage could be detected in nerves at rest. Some twenty years later Julius Bernstein (1868) found that nerve or muscle fibres had an excess of positive ions on the outside and an excess of negative ions on the inside. The Italian physiologist Carlo Matteucci had previously shown, in 1842, that an electrical potential existed between the outside surface of a muscle and its inside by cutting the muscle across.

### Plasmolysis

Around about the same time, Karl Nageli and K. Cramer (1855) discovered the phenomena of plasmolysis and deplasmolysis in plant cells. These cells could be made to change their volume by altering the magnitude of the osmotic pressure of the incubating solution. Fluid could be withdrawn or admitted into the cytoplasm depending on the nature of the substance added to the incubating medium. When the osmotic pressure in the bathing fluid was higher than that inside the cell, fluid left the cytoplasm and the cell volume decreased so that the cytoplasmic surface became detached from the thick plant cell wall (plasmolysis). If the substance or solute added to the bathing fluid did not leak into the cell the plasmolysis would continue for a considerable time. If, however, the solute was permeable and quickly entered into the cytoplasm it would rapidly establish the same concentration inside the cell which would increase its osmotic pressure and the fluid would return into the cell, re-establishing its original volume (deplasmolysis). These cells resembled osmometers with semipermeable membranes. This easily viewed shrinking provided a direct method for a number of botanists (Pfeffer, de Vries, Klebs and others) to obtain semiquantitative measurements of the permeability of the protoplasmic surface to solutes.

### The membrane is a barrier

Wilhelm Pfeffer, a German botanist, realized that the cell surface acted as a membrane and even forecast that carriage of some substances across it probably meant that the substance was temporarily bound to a membrane component. In 1890 he clearly formulated the membrane concept into two concise statements, (1) the cell is enclosed by a plasma membrane, and (2) this membrane is a universal barrier to the passage of water and solutes. Today, we are not as dogmatic about the latter statement for we know that many solutes do pass rapidly through the membrane faster than by simple diffusion (see page 74). Pfeffer's membrane concept was supremely simple. It could well be stated that all the modern research has been at pains to undo this elegant simplicity!

### The penetration of substances into cells

The realization that the plasma membrane was the major permeability barrier stimulated many studies on the penetration of substances into cells. Although early electrical measurements indicated that the cell membrane had differential permeability to a number of ions, it was the penetration of non-electrolytes that attracted most of the attention. A part of the early work was done on large plant cells but a few pioneer experiments by Grijn and Hedin in 1896–97 were undertaken on red blood cells. They were interested only in what type of substances penetrated into the cells and recorded that substances of like structure had a similar penetration. Critical studies of the plasma membrane function were placed on a sound experimental footing by the work of Charles Overton of Zurich during 1895–1900.

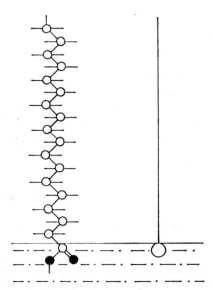

*Fig*. 3. Structural representations of a molecule of stearic acid floating on water. The molecule, like most lipid molecules with polar and non-polar groups, becomes orientated so that the long hydrocarbon chain, with its hydrophobic (water-hating) structures, is in the air phase while the hydrophilic or water-loving carboxyl group is immersed in the aqueous phase. Complex lipids are usually represented diagrammatically as a circular head (the polar hydrophilic part) with an attached straight line (the non-polar, hydrophobic part).

## The lipoidal membrane – Overton's concept

Overton investigated the penetration of alcohols, sugars, bases, amino acids, organic acids, urea and thiourea into plant and animal cells. He found that the introduction of polar groups (hydroxyl, amino or carboxyl) decreased the entry of substances while non-polar groups, like halogens or alkyl groups, increased the entry.

Furthermore, in homologous series of aliphatic alcohols (e.g. methyl, ethyl, propyl, butyl alcohols) or the monocarboxylic acids, as the length of the carbon atom chain increased, so did the penetration of the acid or alcohol into the cell. These results suggested to Overton that the rate of entry of substances into cells was related to the surface barrier of the cell being mainly lipoidal. The word "lipoidal" was his own invention, he used it to indicate lipids and fatty substances in general. This basic idea of a lipid outer membrane being the main barrier to diffusion between the cell and its environment received confirmation in 1933 from the work of the Finns, Collander and Barlund who also studied the penetration of large numbers of substances into plant cells.

## The Gorter–Grendel experiment

Before their publication, however, two scientists, Evert Gorter and F. Grendel, working at the University of Leiden in Holland in 1925, had devised an experiment of classic simplicity to investigate whether the animal cell really could be surrounded by a lipoidal membrane. The red blood cell was used, because they needed a cell population that was homogeneous, could be obtained in large quantities and easily counted, and had a structure that would allow an accurate estimation of its total surface area. Moreover, as the red cell had no nucleus and contained practically nothing else but haemoglobin, any lipid extracted would come from the plasma membrane. All they had to do was to obtain some red blood cells, extract them with a lipid solvent (acetone) and measure the surface area the extracted lipids would have if they were a continuous film one lipid molecule thick. By comparing this area with that of the surface of the red cells, an estimate of the number of lipid layers that could be contained in the plasma membrane would be possible. The area of the lipid films was obtained by using the techniques of the physical chemist Irving Langmuir. He had invented a special balance in 1917 that could measure the area of single molecular layers of lipid.

## The Langmuir film balance

Langmuir developed this balance to study films of lipid that were floated on the surface of water. These monomolecular layers could be made because lipid molecules orientated themselves at the air/ water interface with their charged or polar group at one end in the water and their non-polar or carbon chains sticking out in the air

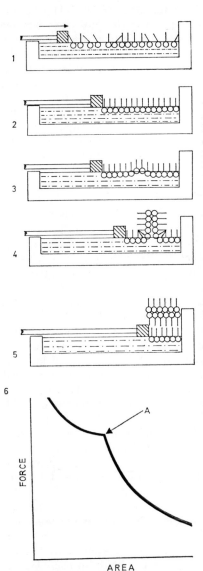

*Fig.* 4. The basic concept behind the Langmuir trough. A known amount of lipid is placed on a Langmuir water trough. The molecules form a film but are orientated randomly. All, however, have their hydrophilic heads in the water and their hydrophobic tails in the air (1). As the float is pushed along the trough to compress the film the molecules are compacted together (2). When the pressure on the film is such that all the molecules are in contact with one another and have formed a perfect monomolecular film further advance of the plunger makes the molecules of the film slide over one another (3). Suddenly they break away from the surface (4) and fold over to form a series of lipid layers (5). It is this sudden change in pressure on the plunger between stages 3 and 4 (shown as A in the graph (6)) that denotes when the lipid film goes from a compact monomolecular film to a series layer. This unique transition point allows one to calculate the surface area of the lipid when spread as a monomolecular layer. If the weight of lipid and molecular weight are known the volume occupied by each molecule can be estimated.

(Fig. 3). In a large space some lie on the surface (Fig. 4, 1) but as the area is reduced by compressing the film the molecules become tightly packed and arrange themselves in vertical orientation (Fig. 4, 2). Langmuir's balance simultaneously measures the area of the film and the pressure that it exerts on a moving mica float that pushes against the edge of the film. Compression is increased until the pressure remains constant or falls. This indicates that the film has collapsed (Fig. 4, 5). Knowing the weight of lipid applied on to the water surface of the balance, the volume occupied and the pressure it exerted before it collapses as a monolayer, it is possible to calculate the total surface area of any amount of the same lipid spread out as a monolayer.

## The data of Gorter and Grendel

Gorter and Grendel counted the number of red cells per millilitre of human blood (4,740,000). They calculated the surface area of one red cell assuming the structure was a disc. The surface area was thus twice the square of the diameter ($2d^2$) or 99·4 $\mu^2$. By multiplying the total number of red cells by the surface area of one, they obtained the total surface area of all the red cells in the 1 ml. of blood. This was equal to 0·47 sq. metres. The surface occupied by all the lipid extracted from 1 ml. of blood (as measured on the Langmuir balance) was equal to a monolayer of 0·92 sq. metres. Thus, dividing the total surface area of the cells into this figure gave approximately 2. They concluded that "the quantity of lipids is exactly sufficient to cover the total surface of the chromocytes (red blood cells) in a layer that is two molecules thick". They therefore supposed that the red cell was surrounded by a bilayer of lipid, polar groups directed to the inside and outside (Fig. 5). This was the first statement of the now almost universally accepted idea that the cell membrane is basically composed of a bimolecular lipid leaflet, yet at the time it attracted few advocates and appeared to have been either ignored, overlooked or criticized. Indeed the data of Gorter and Grendel came in for heavy criticism. Their calculation of red cell area was obtained from dried films of cells and was clearly unphysiological. The assumptions that all the lipid was extracted by only one type of solvent (acetone) and that no lipid was present in the cytoplasm were thought inaccurate while their presumption that the surface area of the lipids measured on a Langmuir balance would be the same as that in the membrane in the living state was invalid. Although these criticisms are to a large extent well-founded, it should never be forgotten that Gorter and Grendel's bimolecular lipid leaflet was a milestone in our understanding of the cell membrane. It became the seminal idea for later theories about membrane structure. Even using modern data with

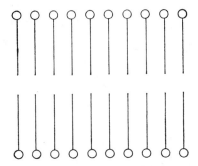

*Fig.* 5. The classic lipid bilayer membrane surrounding the red cell that was proposed by Gorter and Grendel in 1925. The small circles represent the hydrophilic groups of the lipid. The attached lines their hydrophobic groups.

the fewest possible assumptions, the ratio of the red cell's lipid to its surface area is approximately 1·56 (Korn 1966). Until other cells are studied by the same procedure we cannot say whether this lower figure is significantly different from the magic 2 or not.

### The heterogenous lipid membrane of Collander and Barlund

The majority of the early studies published on the entry of substances into plant cells was of a qualitative rather than quantitative nature. Many studies assumed that disappearance of the substance from the medium in which the cell was incubated could be directly equated with its entry into the cell cytoplasm. If substances are metabolized or are bound on the outside of the cell this equation is not correct. Two Finns, Paul Collander and H. Barlund (1933) published the outstanding quantitative study of the period. Using cells of Chara ceratophylla they estimated the permeability of the cells to a host of substances by measuring the actual amount of substance that penetrated into the cell sap. Their measure of permeability was indeed one of true transfer across the membrane into the cell. They published their results in a strikingly simple graphic manner: the permeability of the substance, P, (mol/cm$^2$/hour/mol per litre) was plotted against its oil–water partition coefficient (Fig. 6). At the same time a measure of the volume of the various molecules was calculated for each substance used. From the graph on

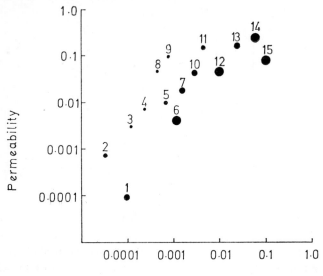

Partition coefficient (olive oil/water

*Fig.* 6. The permeability to various non-electrolytes of the plant cell Chara plotted against the olive oil–water partition coefficient of these substances. The molecular volume of the various substances is represented by the size of the solid circles. This way of plotting the data was instituted by Collander and Barlund in 1933. Increasing lipid solubility leads to increasing permeability even if the molecules are fairly large. Molecules with poor lipid solubility (low oil–water partition coefficients) can enter reasonable fast if they are of small size. The numbered circles on the graph refer to the following compounds; 1 = malonamide, 2 = glycerol, 3 = urea, 4 = methylurea, 5 = thiourea, 6 = diethylmalonamide, 7 = ethylurea, 8 = acetamide, 9 = formamide, 10 = diacetin, 11 = propionamide, 12 = diethylurea, 13 = valeramide, 14 = trimethyl citrate.

which all three measurements were plotted (the molecular volume of the substance denoted by the area of its circle) Collander and his workers concluded that "there is clearly a somewhat close concordance between the oil–water solubility of substances on the one hand and their permeability on the other; this is not merely a general concordance but at least approximately a direct proportionality". This line of evidence was in direct accord with Overton's experiments and conclusions. The important extra fact that Collander proposed was based on the molecular volumes of substances used, for he also found that "the smallest molecules obviously permeated faster than would be expected on account of their oil solubility alone".

The hydrophilic substances that diffused faster than expected from their oil–water partition coefficients were water, formamide, methanol and ethylene glycol. To explain their rapid entry into the cell, Collander proposed that the membrane was not completely homogeneous. Although it acted as a lipid membrane there had to be areas that allowed small hydrophilic molecules to penetrate quickly. In Collander's words "the medium-sized and large molecules penetrate the plasma membrane only when dissolved in the lipids, the smallest molecule can also penetrate in some other way. Thus the plasma membrane seems to act both as a selective solvent and as a molecular sieve." The cell membrane appeared to be a mosaic containing lipid areas and aqueous pores.

### Re-assessment of the experiments of Collander by Danielli

Collander and Barlund had arrived at their conclusions that the membrane behaved as a mosaic of lipid and aqueous pore areas by their ingenious method of plotting the permeability directly against the oil–water distribution ratio. At the time of their experiments, everyone assumed that the diffusion of a highly lipid–soluble molecule across a biological membrane would be similar to that of a poorly lipid–soluble molecule.

Some ten years later, James Danielli, working at Cambridge University, realized that a diffusing molecule had three sites of resistance when penetrating a lipid membrane separating one aqueous phase from another. These sites were (1) the water–lipid interface, (2) the interior of the membrane, (3) the lipid–water interface. Previously everyone had thought that 1 and 3 were very much less important than 2. Danielli showed that this was in fact not so, for in thin membranes the effective resistance to free diffusion is almost entirely at the membrane–water interface. For different molecules the cell membrane will present different resistances. A molecule that penetrates easily because of its high lipid solubility (say benzene or propane) finds no difficulty in diffusing at 1 or 2

but at 3 the substance has to have enough energy to leave the preferred lipid phase and jump into the aqueous phase. Similarly, with
a molecule that is not very lipid soluble (like glycerol) the difficulties
this substance will encounter diffusing across the membrane are at
1 and 2. It will not find 3 any difficulty as it prefers going from a
lipid phase into an aqueous one. Analysis of these two types of
diffusion revealed that for a homogeneous lipid membrane a simple
linear relationship between the permeability coefficient, P, and the
partition coefficient, B, could not be expected for both slowly and
quickly diffusing substances. Using mathematical and theoretical
arguments too advanced for inclusion in this text (they filled six
pages in the original book!) he showed that for slowly penetrating
molecules $P = \frac{1}{2} aB$, where a varies with the square root of the molecular weight (M) but for rapidly penetrating molecules $P = \frac{BK}{\sqrt{M}}$
K being a constant involving a number of complex terms. When
Collander's data was plotted by this more sophisticated approach,
Danielli found that all the substances used fitted the graphs as if
they were penetrating a homogeneous lipid barrier, and gave no
indication of aqueous pores.

### Aqueous pores in a membrane

Although the analysis of Danielli showed that it was not possible
to use Collander's data to prove the existence of pores in the lipid
membrane of Chara ceratophylla, another early study on the
bacterium, Beggiaitoa mirabilis, appears free of any interpretive
errors. Ruhland and Hoffman, in 1925, used this filamentous,
sulphur bacteria to measure the penetration of a large number of
non-electrolytes by the plasmolytic method previously described.
The permeability of various compounds tested bore little relation
to their ether–water partition coefficients, but they appeared to be
related to molecular volume. The membrane allowed molecules to
penetrate according to their diameter rather than to their lipid
solubility, indicating a sieve-like structure rather than a lipid layer.
The molecular size of the "pore" was about the diameter of the
sucrose molecule (approximately 8·8 Å).

### Summary of the membrane structure as inferred from early functional studies

In the previous narrative, a chronological sequence of the main
experimental studies on the plasma membrane was presented. It

will be useful to collect the ideas about the cell membrane inferred from these experiments. Five basic facts can be presented as a summary.

1. The membrane is a universal barrier to passage of solutes (Pfeffer's postulation).

2. It is a thin layer of lipoidal material (Overton's concept).

3. The lipoidal layer is a bimolecular leaflet with the lipid polar groups facing outwards (Gorter and Grendel's conclusion).

4. It could have aqueous pores, approximately 8 Å in diameter (Ruhland and Hoffman's data).

5. Different membranes have mosaic structures with varying proportions of lipid areas and pores (Collander's hypothesis).

# CHAPTER 3

# The Paucimolecular Membrane Model

The next significant development of our ideas on the structure of the cell membrane came from the partnership of James Danielli of Princeton University and Hugh Davson of the University College of London. Their model differed from the previous one in that it explained some physical measurements that were being made on the surface of cells.

### The paucimolecular membrane of Danielli and Davson

Studies of animal eggs had led a number of people to believe that naked lipid could not be present on their external surface. The surface tension of the cells appeared to be significantly lower (0·2 dynes cm.) than that expected if a naked lipid layer was present (about 10 dynes cm.). These experiments by Cole in 1932 and Harvey and Shapiro in 1934 indicated that the Gorter–Grendel bilipid membrane appeared inadequate to explain the low surface tension of cells. Then in 1934 Danielli published two historic papers which gave a possible explanation for the low surface tensions observed. The first paper, co-authored with Newton Harvey, suggested that when protein was adsorbed on to a lipid film the surface tension of the lipid was reduced. The second paper entitled "A contribution to the theory of permeability of thin film" was co-authored with Hugh Davson. In this paper they argued that if layers of protein were adsorbed on to the lipid membrane of cells the measured, low surface tension of cell membranes could be explained. They illustrated the paper with a diagram of a cell membrane (Fig. 7). This Danielli–Davson model membrane was later called the "paucimolecular model" a term suggested by Danielli's previous co-worker Newton Harvey to describe a layer a few molecules thick. At the time, the actual number of lipid molecules in the membrane was not specified, the bimolecular lipid barrier of Gorter and Grendel being completely ignored! The protein–lipid–protein sandwich became the basic model of the plasma membrane and to a large degree has remained so up to the present day. There have been a number of modifications brought about by recent research, but without doubt this molecular arrangement has become one of the oldest theories of molecular biology.

EXTERIOR

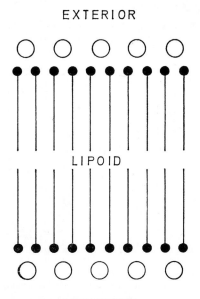

INTERIOR

*Fig*. 7. The paucimolecular model of the cell membrane as envisaged by Danielli and Davson in 1935. The model structure consisted of a thin layer of lipid (lipoid), probably, but not necessarily, of bilayer dimensions, upon which were adsorbed protein films (large open circles). The proteins were originally thought to be globular hence their circular structure.

Danielli later demonstrated that many globular proteins when in contact with lipid films unravelled themselves and formed a denatured layer along the surface about 5–10 Å thick. The denaturing, he thought, would make the protein–lipid film "insoluble" like normal cell membranes, and would also impart to the membrane's structure a solidarity which lipids alone could not achieve. Recent studies on the interaction between proteins and lipid surfaces have shown that Danielli's idea that the first layers of protein were unravelled and denatured by the lipid films is incorrect and that all the protein is in the normal globular form.

## Aqueous pores in the paucimolecular membrane

At the time of the postulation of their model, Danielli could not understand how aqueous pores could exist in their paucimolecular membrane, because he thought that once a pore was formed the surface tension of the lipid layers would keep enlarging it until it destroyed the whole structure. Thus, although the Danielli–Davson model explained both the lipid nature of the permeability barrier

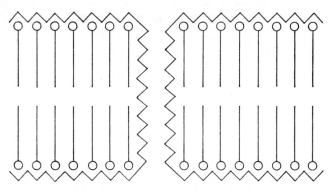

*Fig.* 8. The slit pore model for biological membranes (after Danielli 1959). The bilipid layer is covered with protein (zigzag lines) both on its inside and outside aspects. The slit pore is lined with protein that is continuous with the outer and inner coverings. Such a pore has structural stability and would allow the passage of water and hydrophilic small molecules through the bilipid layer.

of the cell membrane and its low surface tension, it did not satisfactorily explain the rapid movement of small hydrophilic molecules into cells. Much later, Danielli realized that stable pores could be maintained in the bimolecular lipid leaflet if they too were protein lined. His protein slit-pore (Fig. 8) became as well established as the previous model membrane. The pore is stable because the hydrophobic groups of the proteins face into the lipid layer while their hydrophilic ones face out. It was thus kept from being pulled apart by the attraction of the two proteins for one another. Such a pore would allow water molecules and small molecules to pass through the lipid bilayers easily.

## Electrical measurements supporting the membrane concept

Although the early physiologists like Luigi Galvani, Alessandro Volta, Carlo Matteuci and Emil Du Bois Reymond clearly established that biological tissues generated electrical potentials and that changes in these potentials were involved when muscle and nerve became active, progress in the field of electrophysiology was slow until the end of the 1914–18 war. The delay was due to a large extent to the inadequate methods of measuring the very small and rapidly changing potentials across biological membranes. After the war the use of electronic valves to amplify small electrical signals had an enormous impact on electrophysiology. Then when the cathode ray tube became available, accurate measurements faithfully recording the rapid changes of cell potential were possible. Some idea of the rapidity of these changes is shown by the fact that in mammalian nerve, the action potential rises to a peak and returns to rest in less than 1 millisecond ($\frac{1}{1000}$ second)! These variations in potential were then correlated with changes in the permeability of the membrane by using radioactive ions of sodium, potassium and chloride. Even though it was the electronic and chemical tools that made accurate measurement possible, the use of the giant nerve fibres of the squid Loligo greatly simplified the task. This squid has nerve fibres large enough (about 1 mm. in diameter) to allow electrodes to be pushed into the axoplasm without damaging the surface membrane and thus allows the potential difference across the membrane to be directly measured. Few other cells, however, were as large as the Loligo axons so that the electrical activity across many membranes could not be directly measured. What was needed was a minute electrode that could be pushed into cells without causing too much damage. Such electrodes were manufactured at the University of Chicago in 1949 by two physiologists, namely, Gilbert Ning Ling and his supervisor Ralph Gerard.

**The glass microelectrode.** Ling and Gerard heated glass tubing and pulled it into very fine capillaries with tip diameters of less than 0·5 $\mu$. When filled with a conducting electrolyte (usually a concentrated solution of KCl) the glass capillaries became microelectrodes. The electrolyte solution inside the microlectrode was electrically connected to one side of a potential measuring instrument while the other side of the meter was connected to another microelectrode. If the two microelectrodes were well made only a very small potential difference existed between them when they were both inserted into an electrolyte solution. When a muscle cell was impaled by one of the KCl filled microelectrodes, the tip could be pushed right through the membrane into the cytoplasm of the cell. This allowed the potential across the membrane to be recorded

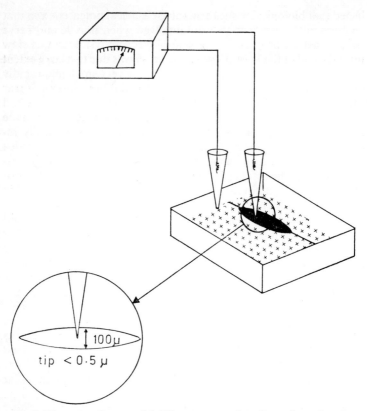

*Fig.* 9. Measuring the potential difference across the cell membrane by micro-electrodes. A piece of frog muscle is suspended in a saline solution that will keep it alive for many hours. Two microelectrodes are shown, one is placed in the saline outside the membrane, the other is allowed to penetrate a fibre until a sharp deflection of the potential is seen on the potential measuring device (either an oscilloscope or millivoltmeter). If this deflection is maintained for some minutes it is assumed to be a measure of the membrane potential. The inset diagram shows a magnified view of a single muscle fibre with dimensions of electrode tip for comparison.

on the voltmeter (Fig. 9). In many cells the electrode could be left inside giving a steady record of a resting potential of about 100 mV for many minutes. This showed that the electrode did not grossly damage the cell membrane. In fact in large cells the effect of pushing a microelectrode into the membrane can be measured by monitoring the p.d. across the membrane with a second microelectrode previously inserted. When this was done, the recording of the transmembrane potential by the first was not affected by pushing in the second electrode. This strongly indicates that the cell damage caused by microelectrodes is not too serious as long as the electrode tips are small. If the experiment is repeated with large tipped electrodes ($>0\cdot5\,\mu$), the damage to the membrane becomes so extensive that the cell becomes permanently injured and little electrical activity can be measured across its membrane. The measurement of transmembrane potentials by intracellular electrodes gave, of course, a powerful boost to the cell membrane concept, but more important than this it revolutionized electrophysiology. At last it was possible to measure what happened inside nerve and muscle cells during their normal activity.

A facet of the cell membrane that the microelectrode work discloses, apart from electrical activity, is its dynamic property of self-sealing or healing small discontinuities. When a microelectrode is pushed into a cell and then pulled out the local damage to the membrane is repaired. Re-penetration of the membrane with the electrode will record the same electrical activity as that of a fresh cell. This property of membranes to repair themselves is an excellent reminder that the membrane is truly a living barrier.

### The electrical resistance of the cell membrane

In the late 1930s and early 1940s a number of workers found that while the cytoplasm of cells was a good conductor of electricity the membrane was not. The electrical resistance of the membrane was similar to that of a thin layer of lipid. The measurements of the resistance and capacitance of nerve and muscle membranes are shown in Table 2. The membrane of the axon, like that of a number of cell membranes, has an electrical capacity of $1\,\mu$ Farad/$cm^2$. Knowing that the dielectric constants for lipids and oils are between 2 and 7, it is possible to calculate the thickness of lipid needed to give this axon membrane a capacity of $1\,\mu$ F/$cm^2$. We can treat the membrane as a simple condenser and apply to it the standard electrical formula for capacity. $C = \dfrac{1\cdot1\,K}{4\,\pi\,d}$ where $K$ = dielectric constant, $d$ = thickness of membrane in cm., $\pi$ is the constant 22/7 and $C$ is in pico-Farads or $10^{-12}$ Farads. Assuming a $K$ of about $5\cdot7$ for the lipid, the thickness of this lipid dielectric layer would be approximately 50 Å. Table 2 also shows measurements of the conductivity of the axon and muscle membrane, as mentioned previously it is surprisingly low, or in other words the membrane resistance is remarkably high. Assuming its lipid thickness is 50Å and the specific resistance about $10^{10}$ ohms cm. it is something like eight orders of magnitude higher than that of the axoplasm in the nerve fibre or of the outside solution. Furthermore, there is normally an electrical potential difference across this membrane layer amounting to some 100 millivolts (mV), the inside negative to the

TABLE 2. *Electrical properties of nerve and muscle cells*

|  | Nerve (squid axon) | Muscle (frog) |
|---|---|---|
| Diameter ($\mu$) | 500 | 75 |
| Membrane resistance (ohm/cm²) | 700 | 4,000 |
| Membrane capacity ($\mu$Farad/cm²) | 1 | 2·5 |
| Resistivity cell cytoplasm (ohm/cm) | 30 | 200 |
| Resistivity extracellular fluid (ohm/cm) | 22 | 87 |

outside. This does not seem very large until the potential gradient is converted into volts cm$^{-1}$, it then becomes a startling 200,000 volts cm$^{-1}$. This potential across the axon membrane can be increased by placing an electrode on the inside and outside of a giant axon and passing a current through the membrane to augment the natural potential difference, a process known as hyper-polarization. If this is done the membrane structure breaks down when the increase in voltage is between 2 and 4 times normal, i.e. 200 to 400 mV. This is just about the same voltage that breaks down the structure of lipid layers. At this level of potential the electrical field strengths are in the order of 3–6 $\times$ 10$^5$ volts cm$^{-1}$. Thus a delicate lipid layer, a mere 50Å in thickness, can have a dielectric strength exceeding that of either polyethylene or porcelain, two of our best insulators.

Although these simple calculations illustrate the remarkable electrical properties of thin, lipid membranes they can only give us an idea of the minimum thickness of the membrane. Moreover, this calculated thickness will only be that of the lipid while the pauci-molecular model has at least two further layers of bound protein to take into account.

**The thickness of the red cell membrane estimated from chemical and physical data.** An estimate of the minimum thickness of the red cell membrane can be obtained from the data of Gorter and Grendel. Their figures for the lipid content of red blood cells would give a membrane thickness of about 30 Å. This value would be in good agreement with that obtained by Fricke and Curtis in 1935 who used electrical measurements of the impedance (resistance) of red cells and estimated that a 40 Å thick lipid layer would account for their measured value. More direct methods have been used to measure the thickness of the membrane of red cells by preparing red cells with little or no haemoglobin inside.

**Preparation of haemolysed red cells for membrane studies.** When red cells are placed in solutions of low tonicity, nearly all the haemoglobin escapes from the interior through the membrane. This process is known as haemolysis. It is accepted that the insoluble part of the cell that is left after haemolysis is practically entirely derived from the membrane. These post-haemolytic membranes have been called stroma, stromata, and "ghosts". The latter name is the one in current use.

When ghosts are prepared the total dry weight of material in these membranes can be measured. By assuming a value for the density of the protein and lipid components and calculating the surface area of the ghosts, the thickness of the membrane can be estimated. Fricke, Parker and Ponder did this experiment in 1939 and calculated a thickness of 120 Å. Other estimations on red cells have since been undertaken giving values ranging from 52 Å up to 125 Å. Part of this large variation is dependent on the manner of haemolysis and part on the amount of washing that the prepared ghosts receive. The more the ghosts are washed, the less lipid they contain. One important fact about these studies remains to be mentioned, namely that all the estimates of thickness are of the dry cell membrane. No one knows how thick the membrane would be when it is hydrated – as it surely would be under normal conditions!

# CHAPTER 4

# Chemical and Ultrastructural Studies of Cell Membranes

## The visualization of membranes by electron microscopy

Even with the finest of modern light microscopes the plasma membrane cannot be seen with light waves for it is simply too small to be resolved. The external shape of cells can be made out under the light microscope, not because the membrane is observed, but because the refractive index of the cytoplasm is different from the fluid outside. This change in refractive index causes a "mirage membrane" at the edge of the cell. The development of the electron microscope between 1930 and 1950 allowed the visualization of cellular structures smaller than the wavelength of light. An idea of the tremendous resolving power of this instrument is obtained when it is compared with that of the eye and the light microscope. The useful limit of resolution of the unaided human eye is about 100 $\mu$ (0·1 mm), that of the best compound light microscope 0·2 $\mu$ while that of the electron microscope is some 0·0001 $\mu$. Today, using the latest electron microscopes, we can resolve biological structures 5–10 Å apart, the size in fact, of quite small molecules! Before we discuss the electron microscope pictures of the cell and their interpretation we must review not only the process by which these pictures are obtained, but also the chemistry of the membrane.

## Fixing and staining cells for the electron microscope

The instrument works by blasting a beam of electrons through a thin section of fixed tissue embedded in a transparent plastic and mounted in a vacuum. Areas that are dense will scatter more electrons than those that are less dense. This gives a picture of contrast on the viewing screen or photographic plate (Fig. 10). Unfortunately, the changes in density between different parts of the unstained cell are too small, with the usual electron beam strengths, to cause efficient scattering. Very powerful electron beams are necessary to examine natural unstained tissue. To obtain good contrast the tissues have to be fixed and positively stained with substances that will effectively absorb on to or react with cell components and increase their mass per unit area, thus giving them more scattering power. The fixatives and stains used for electron

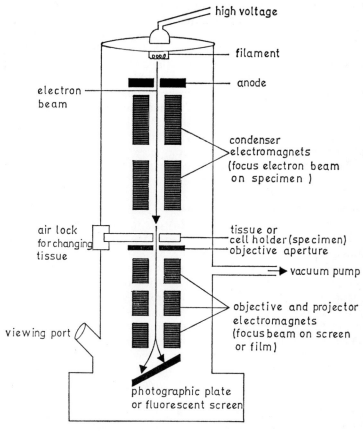

*Fig.* 10. Schematic diagram of an electron microscope.

microscopy are compounds of heavy metal ions which cause good scattering of the electrons. Those most used are osmium tetroxide ($OsO_4$) and potassium permanganate ($KMnO_4$), but uranyl and lead acetate are occasionally employed. Some cellular structures are more likely to react with these reagents than others depending on their chemical groups. The areas of the cell that do react, absorb. or bind the largest amount of heavy metal ions will scatter the most electrons and appear as a dark area in the electron micrograph pictures. Generally, although as we shall see there are exceptions, electron microscope pictures of cells treated with heavy metal salts show a plasma membrane composed of two dark bands separated by a light or clear band, the whole trilaminar or "tramline" structure having a thickness of between 75–100 Å (Fig. 11). Clearly, a critical interpretation of these light and dark areas in cells must depend on our knowledge of the chemical constituents of cell structures, of the possible chemical reactions of these heavy metal fixatives and stains, the nature of the reaction products formed and the way these are deposited in the cell. It is unfortunately just in these areas that our knowledge is most unsatisfactory and nearly all these problems remained unsolved.

The first requirement is to investigate the chemical compounds that are present in cell membranes.

## Chemical constituents of membranes

**Isolation of pure membranes.** (*a*) *Ghosts.* Before a meaningful chemical analysis of a cell membrane can take place, the membrane must be isolated in a pure state, i.e. containing only plasma membrane component with no other parts of the cell. This is far from easy. It is virtually impossible to know how deep into the cytoplasm the living membrane extends and there are also the technical difficulties of preparing isolated membranes, demonstrating that they are in fact surface membranes and estimating their purity. The choice of cell from which the isolated membranes are prepared obviously plays a large part in obtaining preparations with high purity. Red blood cells are again high on the list because of their ease of collection, large surface to volume ratio and absence of intracellular structures. All that is needed is to haemolyse the cells under the right conditions and practically pure membranes are obtained. It is the choice of the right conditions that causes the trouble. Nearly every experimenter working with red cells has developed his particular way of haemolysis that he regards as giving the best "pure" preparation of membrane. Some 2–3 per cent of the total haemoglobin is left in ghosts washed with neutral water, but if acid or alkali is added all this haemoglobin can be removed. A number of people argue that the 2–3 per cent haemoglobin is part of the fabric of the membrane and that its removal by acid or alkali represents a breaking of structural bonds in the membrane. This argument is not unimportant as estimates of the membrane thickness, weight and protein will clearly be affected by the residual haemoglobin.

(*b*) *Other cells*. Only in the last few years have isolated membranes from liver and tumour cells been successfully prepared. They can be obtained by disrupting the cells with ultrasonic vibration and then

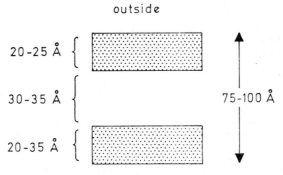

*Fig*. 11. The electron microscopic appearance of the cell membrane after staining with a heavy metal. The membrane usually appears triple layered; two dark bands approximately 20 Å thick are separated by a 35 Å thick clear layer. There is much variation however in the thickness of the various layers in different cells and membranes and with different techniques of preparation.

centrifuging the crude membrane extract in a test tube that has sucrose solutions of different concentrations carefully layered one upon the other. By spinning the extract in this "density gradient" tube, different cell components will float at the different levels of sucrose concentrations. These layers can then be individually sucked off and their contents analysed morphologically (usually by the electron microscope) and then chemically. This explanation makes the whole process sound ridiculously easy but the recognition of contaminations with non-plasma membrane elements in the membrane fraction is often exceedingly difficult.

### Analysis of the pure membrane – gross chemical composition

The chemical data obtained from the various membranes isolated shows that the animal cell membrane consists largely of protein and lipid with a small amount of carbohydrate. The proportions of these fractions in relation to one another vary not only with the type of cell, but also with the species of animal and the method of preparation (as previously described with red cell ghost membranes). Generally speaking the ghost membrane contains 20–40 per cent lipid and 80–60 per cent protein with a very small amount of carbohydrate. Liver cells have been estimated to contain 40 per cent lipid and 60 per cent protein with less than 1 per cent carbohydrate. The estimation of the amount of protein in membranes is prone to more errors than that of the lipid because many membrane proteins are soluble in the solutions used to isolate the structures; they can thus be leached out during purification. Something like 30 per cent of the protein in the membrane of rat liver cells can dissolve out in this manner.

TABLE 3. *Structural designation of α and β phospholipids. The phospholipids stem from the parent compound glycerol.* $R =$ *fatty acid groups and* $A =$ *derivative of orthophosphoric acid*

| | | | |
|---|---|---|---|
| Glycerol (Parent compound) | OH<br>$\mid$<br>$CH_2$ —<br>(α) | OH<br>$\mid$<br>CH —<br>(β) | OH<br>$\mid$<br>$CH_2$<br>(α¹) |
| β-phospholipids | R<br>$\mid$<br>CO<br>$\mid$<br>O<br>$\mid$<br>$CH_2$ —<br>(α) | A<br>$\mid$<br>O<br>$\mid$<br>CH —<br>(β) | R<br>$\mid$<br>CO<br>$\mid$<br>O<br>$\mid$<br>$CH_2$<br>(α¹) |
| α-phospholipids | R<br>$\mid$<br>CO<br>$\mid$<br>O<br>$\mid$<br>$CH_2$ —<br>(α) | R<br>$\mid$<br>CO<br>$\mid$<br>O<br>$\mid$<br>CH —<br>(β) | A<br>$\mid$<br>O<br>$\mid$<br>$CH_2$<br>(α¹) |

The figures for protein, lipid and carbohydrate give some idea of the chemical structure of the membrane but are of little use for visualizing its molecular architecture. For this we need to know the specific type of chemicals in each of those three membrane components.

### Lipid classification

Lipids are classified into simple and compound lipids. Simple lipids are esters of fatty acids with various alcohols. When the alcohol is glycerol the products are fats; when other alcohols are present, waxes. Compound lipids are esters of fatty acids containing other groups in addition to those of glycerol and fatty acids. They are usually divided into three main fractions: (1) the phospholipids; (2) the cerebrosides; and (3) other compound lipids such as lipoproteins, sulpholipids and amino-lipids.

The phospholipids are derived from glycerol by esterifying two of the hydroxyl groups with various fatty acids, while the third is esterified with a derivative of orthophosphoric acid (Table 3). If this derivative occupies the middle –OH group it is a $\beta$-phospholipid while if the group is at either end it becomes an $\alpha$-phospholipid. Because these compounds exhibit stereoisomerism, they will turn polarized light either to the left (Laevo- or L-series) or to the right (Dextro- or D-series). Most naturally occurring lipids are of the L-$\alpha$ form. The complexity does not conveniently end here, for the phospholipids are themselves subdivided into at least a further six groups. These groups are listed in Table 4 with their structural formula where appropriate. In the early days the classification was according to their solubility in alcohol. Those that were soluble were called "lecithins" and those that were insoluble were called "cephalins". With further study this distinction became irrational as the cephalin fraction was shown to have substances that also dissolved in alcohol. Lecithin and cephalin have now become terms for more specific structures; lecithin is now regarded as phosphatidyl choline, while cephalin is phosphatidyl ethanolamine.

### The lipids and fatty acids of the red cell membrane

At the present time the only reliable estimations of lipids in plasma membranes seem to be those from red blood cells. Even here a large number of the early studies had very variable data, presumably due to inadequate techniques of separation. We now have new gas chromatographic processes that vaporize the lipids into an inert

TABLE 4. *The major phospholipids in cell membranes. The chemical structure of many phospholipids are complex and as yet there is no simple, satisfactory classification. Those illustrated in this table show a few basic features, i.e. the glycerol backbone with the phosphoric acid group and substituent chemical groups $R_1$, $R_2$ and $R_3$. The compounds are charged and have polar and nonpolar ends to their molecules (see Downs, 1963).*

| Trivial name | Chemical or specific name | Structural groups or identifying groups |
|---|---|---|
| Lecithins | Phosphatidyl choline | $R_3 = -CH_2CH_2N^+(CH_3)_3$ |
| Cephalins { Phosphatidyl serine / Phosphatidyl ethanolamine | $R_3 = -CH_2CHNH_2COO^-$ <br> $R_3 = -CH_2CH_2NH_3$ |
| Lipositols (Phosphoinositides) | Phosphatidyl inositol | $R_3 = $ inositol ring structure (with OH groups) |
| | Phosphatidic acid | $R_3 = H$ |
| Plasmalogens | Acetal phosphatides (suggested that they are called phosphatidal compounds) | aliphatic aldehydes |
| Sphingomyelins | Phosphatidyl sphingomyelins | contain *no* glycerol but have fatty acids, phosphoric acid, choline and complex alcohol |

General glycerophospholipid structure:

$$R_2 - C - O - CH_2$$
$$\quad\quad\quad\; |$$
$$\quad\quad\quad CH - O - C - R_1$$
$$\quad\quad\quad\; |$$
$$\quad\quad\quad CH_2 - O - P - R_3$$
$$\quad\quad\quad\quad\quad\quad |$$
$$\quad\quad\quad\quad\quad\quad O^-$$

Plasmalogen structure:

$$CH_2OCH=CHR_1$$
$$R_2CO-O-CH$$
$$\quad\quad\quad CH_2O-P \begin{cases} O \\ O^- \end{cases} \text{choline or ethanolamine}$$

Sphingomyelin structure:

$$CH_3(CH_2)_{12} CH=CH \; C \overset{OH}{\underset{H}{C}} CHCH \; CH_2-O-P-O-CH_2$$
$$\quad\quad\quad\quad\quad\quad\quad\quad\quad | \quad\quad\quad\quad || \quad\quad | $$
$$\quad\quad\quad\quad\quad\quad\quad\quad\quad NH \quad\quad O \quad\quad CH_2$$
$$\quad\quad\quad\quad\quad\quad\quad\quad\quad C=O \quad\quad\quad\quad N^+(CH_3)_3$$
$$\quad\quad\quad\quad\quad\quad\quad\quad\quad CH_3(CH_2)_{22}$$

(Sphingomyelin)

gas stream that passes through various columns which preferentially adsorb and separate the lipids. Lipids can also be applied to paper treated with silicic acid which again preferentially absorbs the different lipids. They are then eluted with a variety of solvents. These modern techniques allow reproducible results to be obtained. Table 5 is a summary of the recent analysis of human red cell lipids, they are mainly a mixture of cholesterol, lecithin, cephalin and sphingomyelin (the total lipid is some 20–40 per cent of the total membrane). Small amounts of lysolecithin and phosphatidic acid are present, while less than 3 per cent of the total lipids present are compounds like glycerides, cholesterol esters and free fatty acids.

**Fatty acids.** The major fatty acids in ghosts are palmitic, stearic,

TABLE 5. *The major lipid fractions in the red blood cell membrane* (human). *Because there are large variations in the amounts of lipid estimated by different studies, the percentage figures shown are either ranges or approximations taken from a number of published results*

|  | Total lipid % | Classification |
|---|---|---|
| Lecithin (phosphatidyl choline) | 32–39 | Phospholipids |
| Cephalin { phosphatidyl serine / phosphatidyl ethanolamine } | 24–29 | |
| Sphingomyelin | 15–37 | |
| Phosphatidic acid | 1 | |
| Cholesterol (free) | 29 | Neutral lipids, glycolipids |
| Cholesterol (esterified) and free fatty acids | 1·5 | |

TABLE 6. *The major fatty acids in red cell membranes. Stearic and Palmitic have no double bonds, oleic has one, linoleic has two while arachidonic has four. The more double bonds the more unsaturated is the compound and the more likelihood that the molecule exerts strong attractions for other molecules, hence stabilizing the membrane*

| Palmitic acid | $CH_3(CH_2)_{14}$ COOH |
|---|---|
| Stearic acid | $CH_3(CH_2)_{16}$ COOH |
| Oleic acid | $CH_3(CH_2)_{17}$ CH:CH $(CH_2)_7$ COOH |
| Linoleic acid | $CH_3(CH_2)_4$ CH:CH $CH_2$ CH:CH $(CH_2)_7$ COOH |
| Arachidonic acid | $CH_3$ (CH:CH $CH_2$ $CH_2)_4$ $CH_2$ $CH_2$ COOH |

oleic, lineoleic and arachidonic acid (Table 6). Oleic acid (48 per cent of the total) and arachidonic (17 per cent of the total) are the two most abundant unsaturated fatty acids present, with about 23 per cent saturated fatty acids. The unsaturated ones probably stabilize the membrane structure because their olefinic bonds exert stronger attractions on other molecules than do the paraffinic hydrocarbon bonds of the saturated acids.

**Species differences in lipid content.** The lipid data for red cells has been presented as if there was but one composition for all types of cell studied. Unfortunately, this is far from the case. Red cells from

TABLE 7. *The major phosphatide components of mammalian red cell membranes (ghosts). Figures are percentages and are taken from data of de Gier and van Deenan (1961) in Maddy (1966)*

|               | Rat | Rabbit | Pig | Ox | Sheep | Man |
|---------------|-----|--------|-----|----|-------|-----|
| Lecithin      | 56  | 44     | 29  | 7  | 1     | 39  |
| Cephalin      | 18  | 27     | 35  | 32 | 35    | 24  |
| Sphingomyelin | 26  | 29     | 36  | 61 | 63    | 37  |

each animal have their own particular lipid composition. The variations in some of the major phosphatide components of a number of mammalian ghosts are shown in Table 7. The idea has more than once been suggested that such variations in composition might be reflected in variations of permeability.

## The significance of lipid composition to permeability

As stated above, each red cell from each species has its own lipid "fingerprint". In the sequence rat, rabbit, pig, horse, ox and sheep, the ratio of sphingomyelin to lecithin falls. The palmitic and arachidonic acids also decrease but oleic acid content rises. It is striking that in this sequence of animal red cells the permeability to hydrophilic solutes like urea and thiourea also falls from a maximum in the rat through rabbit, pig and horse down to a minimum in ox and sheep. It would be satisfying to be able to equate these changes in permeability with this variation in lipid composition. Preliminary experiments, however, indicate that dietary induced changes that occur in red cell membrane lipids do not correlate with the permeability changes.

Another study of the relation of permeability to the membrane lipid structure was made by Roelefson and his colleagues in 1964. They used the fact that the lipids of the red cell membrane can be simply divided into (*a*) loosely bound lipid (extracted with dry ether), and (*b*) strongly bound lipid (extracted with an alcohol–ether mixture). Species of red cells with high permeability to thiourea (rabbit and man) had more loosely bound lipid than those with a low permeability to thiourea (sheep, ox). Once again it would be nice to link the structural data with the measured variation in permeability, but, as with the previous study, it is too early for us to make such simple correlations.

Although we have been examining the lipids of plasma membranes only, it is instructive to look briefly at nerve myelin. This membrane is formed by Schwann cells wrapping themselves around nerve axons. Its function is to act as an insulator for the nerve, i.e. as a permeability barrier. Because of the need for a "tight membrane" we should expect to find that the myelin has a high lipid content and that the nature of these lipids is somewhat different from more leaky membranes. In fact both these suppositions are correct. Human and animal myelin contain about 76–78 per cent lipid (compared to the lower values of 20–40 per cent for plasma membranes). They also have a large content of cerebroside (about 15 per cent), a lipid that is not contained to any appreciable degree in red cells or mitochrondrial membranes.

We must leave the lipid composition of membranes with the sad thought that although modern research is making us increasingly aware of the large number of lipid structures in each of the old lipid fractions, this added knowledge is not at the moment making it any easier to understand how the lipid is involved with the permeability of the membrane.

### The protein of the red cell membrane

Though there were technical difficulties in studying the lipid part of the cell membrane these are small in comparison with those experienced in the study of its protein. This is so despite the fact that about 80 per cent of the membrane is protein. Our ideas about the protein composition of ghosts are little more than a collection of names for various membrane extracts. Furthermore, the proteins are without doubt altered and denatured during the chemical extractions! When ghosts are treated with alcohol and ether, a denatured protein remains named stromatin. If the ghosts are washed at pH 9 another product is left named stromin. Extraction of this with ether forms yet another substance called elenin. Further extraction with alcohol–ether converts elenin to stromatin. The only really important fact about all this manipulation and extraction is that elenin is a rod-like lipid–carbohydrate–protein complex about 5 $\mu$ long and 1 $\mu$ wide which possess both the A and B blood group substances and the Rh antigen activity of the red cell. It is thus thought that it is on the outside of the cell membrane.

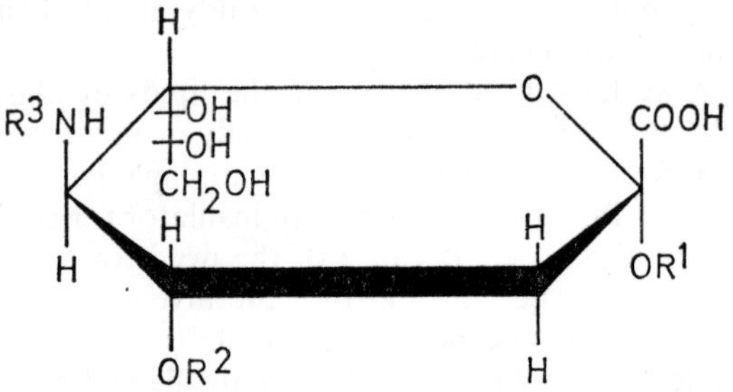

*Fig.* 12. The chemical basis of sialic acids, a feature of the structure of the outer cell surface. When $R^1$, $R^2$ and $R^3$ are all hydrogen ions the acid formed is called neuraminic acid. Naturally occurring neuraminic acids are acylated, $R^1$ and $R^2$ = H but $R^3$ = $CH_3O$. Sialic acid is the group name for all the acylated neuraminic acids.

Although the various fractions of protein that are prepared by the extraction processes could be structural components of the membrane, no one knows for certain whether they are simply artifacts of the various chemical and physical manipulation. Other proteins exist in the cell membrane, but they appear to be present in very small quantities and most are of unknown significance. Some represent metabolic enzymes while others are probably the proteins of the mechanism for transferring ions across the red cell membrane.

**Enzymic degradation of the cell surface protein.** New attempts to study the nature of the proteins on the surface of cells has culminated in the finding during the last few years that at least 90 per cent of the protein of the ghosts contains a substance known as sialic acid (Fig. 12). All of this sialic acid can be split off by an enzyme called neuraminidase, which is itself too large to penetrate into the cell's interior. It is thus assumed that the sialic acid must come from the cell surface, indicating that it is covered with sialoproteins. The sialic acid is probably the source of the major part of the negative charge that resides on the surface of the red cell.

## The concept of the "structural protein" of membranes

In 1963 David Green and his co-workers at Wisconsin extracted mitochondrial and ghost membranes and obtained insoluble proteins which they thought were the basic "structural protein" of each membrane. The main evidence for this idea of structural proteins came from their studies on mitochondria. They found they could remove up to 92 per cent of all the lipid from the mitochondrial membrane and still find a normal structure when it was viewed with the electron microscope. Clearly the lipid in the membrane could not be giving it its "form". They argued that as the lipid was not involved in structure the likelihood was that the proteins of the membrane were. They formed the backbone of the membrane with the lipid attached by hydrophobic bonds. If this was the case, removing the lipid would expose the hydrophobic groups and create a protein insoluble at physiological pH. Their structural proteins were indeed found to be insoluble at physiological pH. Furthermore, these proteins bound phospho-lipids as would be expected if they were normally joined to them in the membrane. Green's theory proposes that all membranes have their own specific structural protein – in fact there would be a "class" of such proteins. One major difficulty is for the experimenter to know that he has obtained one of these "structural" proteins after he has extracted it from a membrane. Might he simply obtain a non-specific protein which was not part of the membrane's backbone? Realizing that this problem of identification existed, Green and his colleagues suggested that any structural protein extracted from membranes must (1) be insoluble in water, (2) bind phospho-lipid and (3) interact with other membrane proteins to form a soluble complex.

The membrane structure proposed by Green and his colleagues cuts dramatically across the accepted structure for normal plasma membranes. The bimolecular lipid leaflet appears to be less dominant; what becomes important is the interaction between the structural proteins upon which are bound phospho-lipids. More details of this new approach to membrane structure will be discussed after reviewing the classical electron microscope pictures of membranes.

## The interaction between lipid and protein in the membrane

If we accept that our knowledge of the lipids in the membrane is unsatisfactory, and that our information about the protein structure is poor, it needs little comment to add that our knowledge about their interaction to form the membrane proper is abysmal. Electrostatic attraction between the charged polar groups of phospho-lipids and proteins and van de Waals forces between their non-polar groups have been suggested as the major forces in the lipid–protein interaction.

The relation of protein to lipid, although poorly understood, is vital for it will clearly affect the reactions of both components to the stains used for electron microscopy. We must now return to these reactions of osmium tetroxide and permanganate with lipids and proteins.

## The reactions of osmium tetroxide and potassium permanganate with cell constituents

Osmium tetroxide was used as a histological reagent to fix and stain cells long before the electron microscope was invented. It stains proteins and lipids containing unsaturated fatty acids black. This early indication that it reacts with both substances obviously makes great difficulty in the interpretation of the electron dense areas of osmium tetroxide treated cells. Are these dark areas lipid or are they protein? They could even possibly be the junctions between the lipid and protein! Initially it would appear simple to answer these questions experimentally by reacting various pure proteins, lipids and lipoproteins in solution with osmium tetroxide; just observe what stains black and we have our clear-cut answer. Unfortunately it just has not worked out that way. Although the experiments have been done by a number of scientists their conclusions are not of great help and have often been conflicting for it has been found (as we would have expected!), that both proteins and lipid layers will stain with osmium tetroxide depending on the conditions and substances used. Thus it is still not known with absolute certainty what the dark areas in the cell are. Moreover, in long molecules like a lipoprotein even if the osmium tetroxide reacted with the lipid part there is no evidence to suggest that the osmium will precipitate at the point of reaction. Indeed some work indicates that there might be a second reaction shifting the site of precipitation up the molecule to the polar end. To add even further to the chemical confusion, the actual product that is precipitated is unknown. Some chemists state that it is the osmium metal, others osmium dioxide while a final group argue that the osmium is chemically bonded to either the proteins or lipids. There is no signed writ from anyone that all three situations might not exist! As is so often the case with humans we crave for simple explanations but in a system so complex as membrane chemistry we are either foolish or naïve to expect them.

Because of the difficulties of interpreting the reaction of osmium tetroxide with cells, attempts were made to use other chemicals to stain components at the electron microscope level. The first electron microscope pictures of tissue fixed and stained with potassium permanganate were published by John Luft in 1956 working at the Department of Anatomy in Harvard. He found that permanganate preserved the fine structure of cells, especially that of the membrane, extremely well. Permanganate was thought to react with tissue components to give a dark brown precipitate of manganese dioxide, which caused the electron scattering. Although permanganate reacts chemically with lipids, it too is a non-specific reagent with an affinity for a wide range of organic substances. Even though its chemical reactiveness with cellular macromolecules is as unknown a quantity as that of osmium tetroxide, it has become extensively used because it yields such fine pictures of membrane. Everyone using it assumes that the better the preservation of structures, the more likely that this is what the living cell is like, a fact that is not always true. Some of the structures described by early histologists are now known to be caused merely by the fixation and staining process, i.e. they were artifacts not found in the living cell. It may be that permanganate causes good "fixation" of membranes for the wrong reasons. Being an ionized substance it enters cells slowly and has to be used in high concentrations to obtain satisfactory penetration. It is possible that these high concentrations destroy the "living" structure of the membrane causing a reorganization to a more ordered pattern. Permanganate is also thought to dissolve a little of the protein covering the lipid of the membrane. This action would also make the lipid more organized. Thus, even if permanganate does not drastically alter the membrane it may certainly change the relation between its lipid and protein.

## The structure of the cell membrane revealed by positive staining and electron microscopy

We can now critically examine the contribution that the electron microscope has made to our ideas about the cell membrane. Its first dramatic impact was that it silenced once and for all the scientific sceptics that said that the plasma membrane was a "mythological beast" inferred from functional studies only and without any anatomical substantiation. They believed that the cell was a gel phase and that the "membrane" was in reality a phase boundary between the gel of protoplasm and the aqueous environment. Even the poorest of electron micrographs of cells showed that the cell cytoplasm in contact with the environment was stained more densely than the rest, and with this "seeing is believing" proof, the no-membrane school were vanquished.

The reward from the electron microscope picture of cells was even greater, however, than a mere confirmation of the anatomical existence of the plasma membrane. An embarrassment of membranes was found in the cytoplasm and around other structures. The cell became a "bag of membranes" instead of the beloved "bag of protoplasm" that so many authors had described.

**A profusion of membranes – part two.** Our previous profusion of membranes was due to the word being used for a number of biological structures (page 10). Our second profusion of membranes developed after the electron microscope revealed the dark "tramlines" of membranes when cells were treated with osmium tetroxide. Membranes were observed not only externally at the border of the cell but also internally in the cytoplasm and around many cell components. Indeed one of the striking features of the electron micrographs was that they showed the once nearly empty cytoplasm to be filled with small granules (the ribosomes, about $150 \text{Å}$ in diameter) and richly endowed with membranous structures. Some of these membranes had the ribosomes attached on to their surface and appeared "rough". The membranous system in the cytoplasm was named the "endoplasmic reticulum" by two American microscopists George Palade and Keith Porter working at the Rockefeller Institute (now University) in New York in the early 1950s.

**Classification of membranes.** A simple classification of the various membranes that osmic acid and the electron microscope exposed is shown in Fig. 13 using a mucosal cell of the small intestine as the typical example. Membranes can be grouped into internal (cytomembranes) and external membranes. The latter include those at the free surface of the cell (plasma membrane), those in apposition where the cells are tightly bound (desmosomes) and those in apposition where the cell is not tightly bound to its neighbour.

Cytomembranes include those around granules, lysosomes, vacuoles and the mitochondria. The membranes around the nucleus and those of the Golgi complex and endoplasmic reticulum would also appear to be immediate candidates for this group, but Palade and Porter proposed that they were possibly merely the extensive infoldings

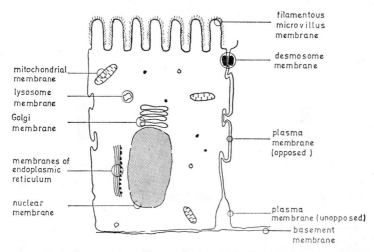

*Fig.* 13. A diagram of a cell *circa* the late 1960s (modelled on the mucosal cell of the small intestine). The electron microscope has uncovered a number of intracellular and extracellular structures unthought of in the previous era of light microscopy. A wealth of membranes invest the cell, both internally and externally. Membranes surround mitochondria, Golgi organs, the endoplasmic reticulum, the nucleus and lysosymes. Externally the plasma membrane can be observed, at the free luminal pole of the cell, to be filamentous while at the joining of cells (desmosome) the membrane substance is condensed. Opposed plasma membranes have a slightly different structure than unopposed ones. The cell rests upon an external membrane, the basement membrane, upon which many epithelial cells stand.

of the plasma membrane. Because the folds are so extensive, when the cell is thinly sliced for the electron microscope, continuity of the membranes with the external plasma membrane is rarely observed. In their idea of the endoplasmic reticulum the cell is like a sponge, the linings of the open channels inside the sponge being the membranes of the reticulum. Such a structure would greatly increase the surface area of the cell. David Robertson supported this concept and put forward a challenging theory that all membranes were of similar construction and that the endoplasmic reticulum evolved because the primitive, simple round cell grew outfoldings which interdigitated (Fig. 14). It is certainly an interesting speculation, but there seems to be definite evidence, as early in fact as 1956, that the various cytomembranes had different structures. This evidence came from the studies of Fritioff Sjöstrand and

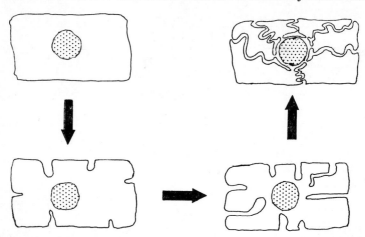

*Fig.* 14. The possible evolution of the organization of the internal membranes known as the endoplasmic reticulum. Initially single cells had the usual simple membrane surrounding the cytoplasm and nucleus (dotted). As the amount of transfer between cells and environment increased due to increasing complexity of cell metabolism the outer plasma membrane became infolded and more tortuous. This allowed a shorter diffusion path from the intracellular to the extracellular fluids and also gave a greater area of membrane for synthetic activity (after David Robertson 1962).

his colleagues. They proposed that there were three basic internal membranes and identified them as α-cytomembranes (membranes associated with granules), β-cytomembranes (those brought about by the infolding of the plasma membrane) and γ-membranes (those of the Golgi complex).

Previously we discussed the difficulties inherent in the interpretation of the reaction of osmium tetroxide and potassium permanganate with even simple chemical classes of biological material. It is therefore not surprising that in the development of electron microscopic examination of stained cells, different people suggested different interpretations for the images they obtained.

This led to rival schools of thought about what were staining with osmium tetroxide and permanganate. Because of this, two major and conflicting ideas about the structure of the plasma membrane developed and went their separate ways even as recently as 1961. In the last few years, however, with the use of tissue fixation before osmium tetroxide staining, there has been some clarification about the differences in interpretation and a measure of agreement has been reached about the fine structure of the plasma membrane.

### Membrane studies of Sjöstrand – images and interpretations

Early pictures of cell membranes obtained with electron microscropes had poor definition because of unsatisfactory techniques for preparing thin sections. In 1953 improved methods allowed ultrathin sections of tissues to be cut, mounted and stained so that electron microscopy could resolve structures very close together.

A major presentation of pictures of cell structure at high resolution was given in the publications of Fritioff Sjöstrand during 1953–54. Cells from liver, kidney and intestine were examined after treatment with osmium tetroxide. The original image of the plasma membrane was of a single 60Å thick dense layer bounding the cytoplasm (Sjöstrand and Rhodin 1953). A clear gap about 110–130Å separated it from its neighbour's 60Å dense layer. The gap was thought to be due to layers of lipid or cement substance. This structure was said to support the Danielli–Davson model. The two dense bands were supposed to correspond to the protein layers and the light areas to the lipid. In retrospect it is clear that the statement that the structure supported the paucimolecular model was not correct. As it represented the apposition of two Danielli–Davson membranes, four dense layers should have been seen not two, unless cell membranes shared their lipid layers and dispensed with their outer protein coats when they become closely apposed. In later years improvements in section cutting, embedding and focusing the electron beam led Sjöstrand to replace the original pictures and interpretation. His present pictures of the plasma membrane are now of a triple layered (tramline) structure, two dense layers separated by a light area. The estimates of its thickness vary and are dependent on the technical procedures used to fix, stain and mount the tissues, but they are within the range 75–100Å. The latest pictures of plasma membranes show a 35Å dense inner cytoplasmic band (thought to be equivalent to the old 60Å single band), a 30Å clear area and a 25Å outer dense band (Fig. 15). With a plasma membrane that was in close apposition to another plasma membrane four densely staining areas were observed as would be expected if two cell membranes of the Danielli–Davson model came close together. The gap between the outer dense bands was some 60 Å.

Sjöstrand's interpretation of the dense staining areas is that the

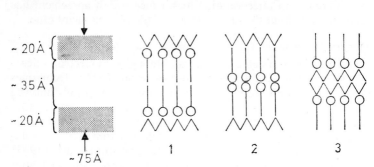

*Fig.* 15. Three possible structures for the unit membrane. The usual electron microscopic image of the plasma membrane stained with potassium permanganate is shown on the left. It consists of a tramline structure of two dark electron dense bands 20 Å thick separated by a clear area some 35 Å thick. Structures 1, 2 and 3 represent the possible arrangement of lipid (circles with tails) and protein (zigzagged lines) that could give rise to such a stained image. In 1, the hydrophobic groups of the lipid are juxtaposed and the hydrophilic groups are covered with protein. In 2 the reverse situation exists while in model 3 the protein layers are covered by the lipid. The usual choice for interpreting the electron microscopic image is model 1 which agrees with the classic Danielli–Davson concept of Fig. 7 (after Robertson, 1966).

osmium binds predominantly to certain amino acid residues in the protein and is subsequently precipitated. These amino acid residues are thought to be hydrophilic as the hydrophobic ones would be deep inside the globular protein molecule and would react only with difficulty with the stain. With regard to the lipid components of the membrane it was proposed that the stain was bound to the polar ends of the phospholipids rather than to the long hydrocarbon side chains. It was assumed that the reduction of the osmium tetroxide by the unsaturated fatty acids results in an unstable binding of the reaction products at the site of double bonds and that there is a migration of the product. This assumption has some experimental backing (see page 42). Finally Sjöstrand advances one further argument to support the protein staining rather than the hydrocarbon chain staining theory. If the structure of a lipo-protein is considered, it is likely that the hydrophobic end of the lipid will be deep inside the structure amongst the hydrophobic side groups of the protein. Thus the more polar phosphate groups of the phospholipid would be located at the surface of the complex. Such an arrangement would favour the stain precipitation at the periphery, where the polar groups are.

### Membrane studies of Robertson – images and interpretation

Another interpretation of the cell membrane structure using the electron microscope and positive staining was proposed by David Robertson working in America. Robertson entered the field of cell membranes by studying nerve myelin in the mid and late fifties. He developed these investigations into a more comprehensive study on the nature of cell membranes which culminated in his challenging idea that all membranes were basically constructed from a unit. In the development of this theory, electron micrographs were not interpreted in the same way as Sjöstrand's. A detailed assessment of all the various experimental pieces of evidence that brought Robertson to suggest the concept of a unit structure is beyond the scope of this book but a brief summary of the highlights will be given.

**The myelin sheath.** Historically the concept developed from studies on nerve myelin. Chemical and physical studies on myelin had been made for many years. It was known to contain large amounts of lipid which could be extracted with solvents leaving protein behind. The insulating myelin sheath of nerve is formed by the Schwann cell wrapping itself around a nerve axon and forming a spiral like a Swiss Roll. This spiral of membranes has been found to be nothing more than a close winding of a double layer of the plasma membrane of the Schwann cell.

**The polarizing microscope.** The German biophysicist, Schmidt, studied the structure of the myelin sheath in the 1930s with an instrument called the polarizing microscope. This microscope passes a beam of polarized light into a structure and then measures the degree to which the light is rotated by the molecular organization of the structure. If tissues are composed of thin molecules with their long axes parallel to one another the beam of polarized light is rotated, the degree of rotation being related to the orientation of the molecule. Myelin, when investigated by this instrument, turned polarized light. It was found that both protein and lipid rotated the polarized light beam but if the lipid was extracted leaving the protein the light was turned in the opposite direction to that turned by the lipid extract alone. Thus the two had to be at right angles to one another. Schmidt's conclusion was that myelin comprised alternate layers of lipid and protein molecules stacked at right angles to one another (Fig. 16). As X-ray diffraction studies revealed

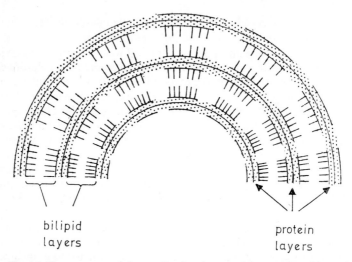

bilipid
layers

protein
layers

*Fig.* 16. The structure of the myelin sheath around nerve axons. The molecular architecture of this structure consists of layers of protein separated by bilipid leaflets. It is like a series of Danielli–Davson model membranes stacked on top of one another. The whole structure acts as a very efficient insulator for nerve and allows rapid conduction of nervous impulses to take place (after Schmidt 1936).

that there was a repeating unit about 170–185 Å thick in the radial direction it was concluded that the unit was a bimolecular lipid layer alternating with protein layers. Further X-ray diffraction studies at much higher resolution by Finean and his colleagues confirmed the earlier results and allowed the general arrangement of lipid and protein layers to be organized in the typical bimolecular layer leaflets with extended protein layers absorbed on to both sides, the thickness of a myelin membrane being about 170 Å (Fig. 16). It was this structure that Robertson used as his guide for interpreting his electron micrographs. The heart of the problem, as always, was what did the dark layers in his electron micrograph pictures of osmium stained myelin represent? Experiments were carried out with artificial lipid membrane model systems to answer the question. Previous observations had shown that structures with no lipid but concentrated amounts of proteins stained densely with osmium tetroxide (i.e. collagen fibres, muscle fibres). Purified lipids, when they were layered, fixed and stained with osmium by the same techniques as used for tissues, gave alternate light and dark bands. The dense bands were separated by twice the length of the lipid molecules. This banding could be explained by the layers being bimolecular lipid leaflets but the dense band could still be either the long hydrocarbon chain or the polar end of the lipid. The logical chemical explanation was that the osmium tetroxide attacked the hydrocarbon chain at double bonds and would be precipitated there. This was in fact what a number of people assumed. Robertson did not accept this conclusion based on classical chemistry and argued that the osmic acid was precipitated at the polar ends of the molecule (as mentioned previously there is now good evidence for this reaction). Robertson's reason for this argument was based on one experimental fact observed with his artificial lipid layers. The layered bilipid leaflets that he had examined could be separated from one another by allowing water to move between them. It was clear that water would go between the polar groups of the lipid layers rather than between the long hydrophobic hydrocarbon chains. When he stained bimolecular lipid leaflets separated by water he obtained a pair of dense lines separated by a light zone. The overall thickness was about 50–60 Å, the distance expected when two lipid molecules met to form a bilayer. As the water must split the bilayers away from one another by moving between their outer hydrophilic polar groups, the two dark outer bands of the separated unit must have been the polar groups of the lipid. Thus osmium tetroxide reacted with lipid to form a dark band at the site of their polar groups. Robertson's early electron microscopic studies of cells were with osmium tetroxide but, disappointingly, the reagent did not give clear pictures of membranes at high resolution. With the demonstration by Luft of the superior fixation by potassium permanganate, Robertson began using permanganate on Schwann cells and axons. The hazy lines that were obtained with osmium staining now showed as a crisp multi-layered structure. The dense line inside the cytoplasm of the cell was about 20 Å thick, then there was a clear gap about 35 Å

thick which in turn was bounded by another thin dense line 20Å
thick, the whole complex being about 75Å wide. Robertson interpreted these layers as being a double layer of the Schwann cell

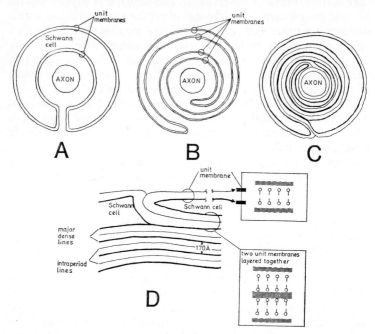

*Fig.* 17. How the myelin membrane is formed around nerve axons. The
Schwann cell spirals around the axon (A). When it has done this a few times
(B) the cytoplasm of the cell disappears leaving the numerous spirals of cell
membranes piled on one another (C). The high-power electron microscope
view of the membranes (D) shows the manner in which the alternating major
dense lines and the interperiod lines are arranged with a repeat every 170Å.
The different thicknesses of these two lines is thought to represent a difference
in chemical content of the inner and outer aspects of the Schwann cell membrane. The inside of the membrane staining more heavily than the outside
(interperiod lines). The possible arrangement of the unit membranes are shown
in the inset diagrams (right-hand side).

plasma membrane. The major dense lines were the insides of the
cytoplasmic membrane while the intraperiod lines were formed by
the outside layers coming together as each layer spirals around the
previous one (Fig. 17).

## Ultrastructural interpretation of different membrane models

Robertson realized that there were at least three possible structures of lipid bilayers with their postulated adsorbed protein coats that were consistent with his staining and electron micrographs. These are shown in Fig. 15 together with the stain image given by permanganate in the electron microscope pictures. Clearly, both 1 and 2 could yield the stain image but 3 was unlikely to give the two dense lines. Structure 2 was unlikely because of the need for the polar heads of the lipids to face outwards to explain the results obtained with water penetration of the lipid bilayers (page 51). Robertson thus settled for structure 1 which in reality was identical with the Danielli–Davson paucimolecular membrane.

**Was the myelin and Schwann cell membrane highly specialized and atypical?** Once the structural implications of the electron micrographs of the Schwann cell and myelin had been formulated, it was necessary to prove that this membrane was not highly specialized and atypical but that its structure applied to cell membranes in general. An extensive survey of different types of membranes in cells from different organs and phyla was carried out. A similar triple layer membrane pattern was found with permanganate treatment, but it was not until the discovery of glutaraldehyde as a tissue fixative for electron microscopy that the same membrane pattern was consistently observed with osmium tetroxide staining. It is now known that osmium tetroxide causes membranes to become excessively leaky, possibly by rearrangement of the lipid–protein. Glutaraldehyde prevents this because it cross links proteins together making them more stable to subsequent staining reactions.

## The unit membrane concept

With the evidence of his electron microscope pictures and his forceful interpretations of their patterns, Robertson became a strong advocate of the unity of membranes in living systems. He proposed that there was a basic "unit membrane" consisting of the lipid bilayer with two protein coats absorbed, the inner cytoplasmic one being of different thickness or composition to the outer. To overcome the considerable criticism that membranes had many different functions and thus could not all be of the same structure, the unit theory was modified by stating that "local variations in the pattern of organization may occur, such as changes in the arrangement of lipids". Many biologists accepted Robertson's unit

membrane as it fitted so neatly with the Danielli–Davson model
but there was a growing body of experiments on mitochondrial
membranes that began to cast doubt on the universality of the
unit membrane.

## The nesting-repeating unit membrane of mitochondria

Every since their discovery in the late nineteenth century mito-

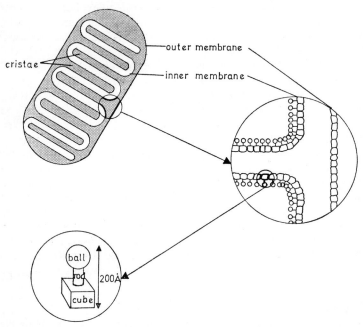

*Fig.* 18. The structure and ultrastructure of the mitochondrion. The mitochondrion is a cocoon-like body with a lattice work of ribs called "cristae".
These consist of an inner and outer membrane (top diagram). The inner membrane is made up of repeating units (inset diagram right). Each of these repeating units consist of a ball-rod-cube unit (inset left) about 200Å in total
length.

chondria have attracted the attention of many biologists. A number
of biochemical studies eventually showed that mitochondria were
the powerhouses of cells, enzymatically converting fuels like glucose
into carbon dioxide, water and energy. The latter was stored in the

form of the phosphorylated compound ATP (adenosine triphosphate). With the advent of high resolution electron microscopy and positive staining the gross structure of the mitochondrion was shown to be a membrane bound bag containing a large number of internal membranes called "cristae" (Fig. 18). It was not, however, until the studies of Fernandez–Moran in 1964, working with David Green's mitochondria study group, that repeating units in the mitochrondrial membrane were first described. These repeating units were found to be composed of two parts, called sectors. In the mitochondrial membrane the unit consisted of a cube-like base piece, a rod-like stalk and a round head (Fig. 18 inset). The stalk and head sector could be detached from the base sector by chemical and physical manipulation and yet still leave the basic membrane structure intact.

### Negative staining of membrane repeating units

Visualization of the repeating sub-units was made possible by high resolution electron microscopy and a technique known as "negative staining". As its title implies negative staining is the opposite of the usual positive staining technique. It was first used at the Cavendish Laboratories in Cambridge in 1959 to examine the structure of virus particles. In negative staining a structure is outlined by being surrounded with an opaque stain but is not actually stained itself (Fig. 19). The stain used is buffered potassium phosphotungstate, the tungstate ion causing electron scattering all around the outlined object so that contrast is obtained, allowing structures approaching 15 Å or less to be resolved. Isolated mitochondria and mitochondria inside cells when treated by this technique clearly showed cube–rod–ball units in the membrane.

In the mitochondria these are believed to be only of one type but other unit structures have been observed in the plasma membranes of liver cells and in the microvilli membrane of the mucosal cells of the small intestine. Although the unit structures look similar they have dissimilar chemical compositions, especially in the case of the ball sector. In the microvilli-ball, digestive enzymes are present (dipeptidase and disaccharidases) while in those from mitochondria the enzyme that breaks down ATP, ATPase, is found.

**The distribution of lipid and protein in the base sector**. Green and his associates found that phospholipid was not randomly distributed in the mitochondrial membrane, the bulk appeared to be with the base sectors and was distributed asymmetrically. When they removed the phospholipid from the base sectors they joined up to form large insoluble units, just as children's small bricks can be pieced together to form a larger cube. If phospholipid was added

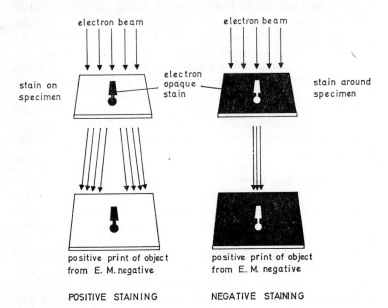

*Fig.* 19. The effects of positive staining and negative staining of a specimen exposed to the electron beam of the electron microscope. In positive staining the electron dense stain attaches itself to the object thus preventing the beam passing through those portions covered with the stain. Wherever the beam passes through and hits the negative the silver grains in the emulsion are reduced. These areas become white when a positive print is made. In the case of negative staining, the opaque stain surrounds and outlines the specimen, thus the electron beam only passes through the outlined object. When a positive print is made of this negative the object is white and the stained surrounds are black. Negative staining can expose details that positive staining obscures.

to these lipid depleted base sectors they did not form these large cubes but formed long chains or sheets joined at their sides. This indicates that the phospholipid covers only two faces of the base cubes and that the other "bare" faces join up with one another by chemical bonding to form a membrane. This "nesting-repeating unit membrane" is presumably the one observed in the mitochondrial, plasma and microvillus membrane with negative staining.

## The nesting-unit membrane and the interpretation of the positive staining micrographs

The previous discussion that membranes contain 50 per cent of a "structural protein" and that it appears to be the major structural foundation of the membrane (page 41) will clearly make a difference in the interpretation of the positive staining of the membrane components by heavy metal ions. The basic membrane sandwich now contains lipid, protein, lipid or perhaps more accurately, "lipoprotein–proteinolipid" molecules. Such a molecular architecture is a direct challenge to the usual accepted paucimolecular membrane of Danielli–Davson and the unit membrane concept of Robertson. It still has, of course, some similarities with these models. For example, there is still the bimolecular lipid leaflet but its spatial array is now somewhat modified.

This nesting-repeating unit membrane of Green can account in simple terms for a number of experimental facts that the paucimolecular model of Danielli–Davson always had difficulty explaining. The variations in permeability of cells to small, hydrophilic solutes can be explained by the possibility of such molecules going between loosely packed base sectors. The permeability of any membrane will thus be a function of its packing tightness. The subtle variations in lipid and protein of red cells could be causing the species variations in permeability to polar substances (page 38).

Enzymatic functions of membranes are explained by the enzyme contents of stalk and ball sectors attached to the base sectors. Variations in enzymatic functions are caused by the variations in the contained enzymes. Specialized mechanisms for the passage of various substances across the membrane, faster than normal diffusion should allow, can also be accounted for by different types of base pieces in the nesting unit scattered through the membrane. Clearly this nesting-repeating unit membrane will create a new approach to biological membranes. However, although there is little doubt that it is the basic structure of the mitochondrial and microvillus membrane there is still much uncertainty as to whether the external plasma membranes of cells in general are built up from such units. Time and more experimental work will tell if it really is a more accurate description of the natural structure of the living membrane. Most scientists feel happier and more convinced about a theory or a set of measurements if another independent person can duplicate or confirm them in his laboratory. This simple rule of independent confirmation of new facts has been of inestimable value. Numerous unconfirmed experiments have often been washed away by what is called "the tide of undiscovery". New facts become rapidly established in the body of science if they can be routinely confirmed. With this in mind it is interesting to find another independent electron microscope study that may have a bearing on the nesting-unit membrane.

### The globular membrane particles of Sjöstrand

The details of the nesting-repeating unit membrane have been studied in mitochondria or in the plasma membranes of small intestinal mucosal cells by negative staining. New electron micrograph pictures of positively stained membranes of the highest resolution (magnifying membranes by 1 million times!) have been published from Sjöstrand's laboratory showing a globular ultrastructure in many of the cytoplasmic membranes. The individual globules are about 5–10 Å in diameter and show up as dark-stained spots separated by less stained regions. The spots are not chemical fixation artifacts as they correspond to similar structures even in membranes that had been fixed simply by freezing rapidly and then sublimating off the ice by vacuum.

The particles were interpreted by Sjöstrand to indicate that the basic structure of the cytoplasmic membrane is of globular proteins around which are lipid molecules, polar groups facing out and nonpolar groups embedded in the protein. Such a structure is remarkably similar to that of the nesting unit base piece.

### The surface of cells

Although we have an approximate idea of the molecular structure of cell membranes we know very little about what their surfaces look like and how their molecules are arranged. There are a few clues, however, from some of the red cell membrane studies that we have previously mentioned. Certainly the blood group AB antigens and the Rhesus factor antigens are present. It is very likely that the rod-like elenin containing these groups is attached parallel to the surface of the cell like plaques. It is probably linked to stromin, the structural protein of the membrane (page 40). Some electron microscope studies using red cells extracted with various solvents suggest that these surface plaques can be removed by such treatment. Finally the surface has a number of negative charges thought to be due to the sialic acid group attached to sialoproteins. Digesting these sialic acid groups off with the enzyme neuraminidase decreases the charge on the surface (page 40).

**The scanning microscope.** Recently there has been developed a new type of electron microscope called the scanning microscope that can give three dimensional pictures of surfaces. It is just beginning to be used to examine membranes of different types of cells. The surfaces of macrophages appear to be somewhat like cauliflowers when visualized by the scanning microscope! Higher magnifications are needed before we will be able to see anything like the detail observed in the conventional electron microscope. A word of warning about the images seen of cell membranes under the scanning microscope is needed. The treatment and preparation of cells for viewing may well cause changes in the structure of the cell membrane so that what we see is undoubtedly not the true form of the surface. The technique is useful, however, when used to substantiate other evidence about the cell surface.

### The use of artificial lipid membrane models as an aid in the study of biological membranes

Very early on in the study of the functions and structures of biological membranes, use was made of the properties of simple layers of lipid and other substances in order to interpret more complex biological measurements. Usually artificial films of phospholipids were made by floating the lipid on to a water surface, as Langmuir did with his film balance (page 15). These monolayers are, of course, between the air–water interface. Objections to this type of film being used as an example of biological membranes often arise because the usual cell membrane separates two aqueous phases. Furthermore the normal biological membrane was supposed to be a bimolecular lipid layer not a monomolecular one. Because of these criticisms, attempts were made in a number of laboratories to make stable bimolecular lipid membranes separating two aqueous phases and to measure their properties. When this was accomplished, a number of surprising facts were found that showed that some of our earlier ideas on the structure of cell membranes inferred from surface tension, light refraction and water permeability measurements on lipid monolayers were in error.

## Artificial bimolecular lipid leaflets

In order to form stable bimolecular phospholipid leaflets separating two aqueous phases Thomas Thompson, at the Johns Hopkins University in Baltimore, dissolved lecithin in n-tetradecane, a water

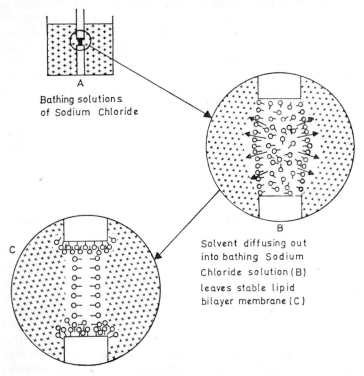

Bathing solutions
of Sodium Chloride

Solvent diffusing out
into bathing Sodium
Chloride solution (B)
leaves stable lipid
bilayer membrane (C)

*Fig.* 20. Preparation of artificial bilipid membranes. Bilipid membranes that mimic the properties of biological membranes can be prepared in the laboratory by applying a small amount of lipid dissolved in a water soluble solvent on to a small hole in a plastic membrane (A). The compartments on either side of the membrane are filled with sodium chloride solution. The water soluble solvent diffuses into the saline (inset B) causing a gradual reduction in the thickness of the lipid film. A large amount of lipid molecules collect at the edge of the film when the film suddenly collapses and becomes a stable bilayer (inset C). The thickness of the film can be measured by optical techniques and has been shown to consist of just two lipid molecules. Their outer circular heads are hydrophilic and their inner straight chains hydrophobic (after Thompson 1964).

soluble lipid solvent. A small droplet of this solution was then introduced into a 2 mm. hole in a thin polythene membrane dividing two chambers filled with 0·1M sodium chloride (Fig. 20). At first the droplet forms a film several microns thick but gradually the water soluble solvent dissolves into the bathing solution and the film gradually gets thinner. As the loss of solvent continues a point is reached when the film abruptly decreases in thickness from about 900Å to 50–70Å. The excess lipid forms small bulges at the edge of the hole (Fig. 20c). As the lecithin molecule is about 35Å in length this film must be two phospholipid molecules thick, assuming the hydrocarbon chains to be fully extended. Small amounts of solvent do remain in the membrane, but this has not been found to be of any importance. The properties of this phospholipid bilayer should approximate to those of cell membranes if a bilipid leaflet is a major structural component of cell membranes. Even the nesting-unit membrane has its lipid layers.

TABLE 8. *A comparison of the properties of artificial bimolecular lipid membranes with those of cell membranes*

| Units | Artificial bimolecular lipid membrane | Cell membrane |
|---|---|---|
| Thickness (angstrom) | $\sim 50 - 70$Å | $\sim 70 - 100$Å |
| Composition | lipid | lipid, protein |
| Electrical resistance (ohm/cm$^2$) | $10^6 - 10^8$ | $10^3 - 10^5$ |
| Electrical capacity ($\mu$ Farad/cm$^2$) | $\sim 1$ | $\sim 1$ |
| Dielectric breakdown voltage (volt/cm) | $\sim 2 - 3 \times 10^5$ | $\sim 3 - 8 \times 10^5$ |
| Refractive index | 1·66 | 1·6 |
| Surface tension (dynes/cm) | 1·0 | 0·1 — 1·0 |
| Water permeability ($\mu$ min$^{-1}$ atm$^{-1}$) | 0·16 | 0·1 — 3·0 |

### Properties of the bimolecular lecithin membrane

A comparison of a number of properties of this artificial membrane with similar measurements made on biological membranes is shown in Table 8. The similarities between the two membranes are striking. Both membranes have high electrical resistances and identical capacitances while their surface tensions and refractive indices are nearly the same. The similarity between their surface tension and refractive indices is very important. Previous measurements of these on biological membranes had been used as proof of the presence of protein on the cell surface because it was found that artificial lipid monolayers had higher values than those found with cell surfaces (page 22). This reasoning cannot now be accepted as the artificial bimolecular lipid layer has no protein adsorbed yet has the same low surface tension and high refractive index as that of normal cell membranes. It is thus possible for naked lipid to be a feature of the normal cell surface.

A property of the artificial bimolecular lipid membrane that is surprising is its very high permeability to water. This is in the same range as that of many normal cells. The usual explanation given of

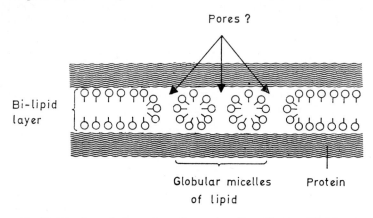

*Fig.* 21. The dynamic formation of pores in cell membranes. Modern concepts favour the possibility that the classic lipid bilayer may undergo reversible changes to globular micelles thus exposing spaces or pores in the membrane. The formation of these pores may be transient, they may flicker along the membrane being "open" only for a short time period. Local alterations in membrane chemistry or ionic make-up may favour their formation or disruption.

the high permeability to water of cell membranes was of continuous aqueous channels or pores in the membrane (page 19). With the artificial lipid membrane continuous pores from one side to the other cannot exist otherwise its electrical resistance (which is very high) would be very low, for such "pores" would allow through movement of the sodium chloride present in the bathing solutions. We must conclude that water can move rapidly through bilipid leaflets even in the absence of pores. How this occurs is not clear.

If we can understand this mechanism in simple artificial membranes we may well be on our way to understanding the transfer of water across the more complex biological ones.

**Various structures formed by lipid layers.** Bimolecular lipid leaflets have been extensively manufactured by people interested in their structure when stained with osmium tetroxide. David Robertson's use of phospholipid membranes to understand the staining of myelin is an example. Other people have been investigating the structure of phospholipid layers when aqueous mixtures of different phospholipids are used. While phospholipids in aqueous solutions are normally thought to exist as bimolecular leaflets they can form other structures or micelles (Fig. 21). Studies have shown that they may exist in a variety of tubular, lamellar or hexagonal arrays. These structures have been shown to be composed of aggregates of globular lipid micelles. It is suggested that the lipids in cell membranes may undergo changes from one form to another, only small perturbations in the chemical environment of the membrane are needed to make such movements possible (Fig. 21). It is because of these new ideas that artificial lipid membranes have become an area of intense research interest and will certainly add further contributions to our understanding of the very much more complex living membrane.

### The function of water in membrane structure

It is perhaps pertinent to add a few words about the role of water in the structure of both artificial and biological membranes. Because the staining and viewing of membranes in the electron microscope needs conditions in which all the water is removed, the function of water in the structure of the membrane was completely overlooked. It is now appreciated, however, that water probably adds a great deal to the molecular organization of cell membranes. Although water is a simple molecule, when fluid it gives rise to highly complex structures. These structures depend on the temperature and the types of molecules that are in its immediate vicinity. When water is in the form of ice, its crystalline-like structure can be fairly well defined and represented because of the relative lack of movement. Various molecules give a varying degree of ice-like structure to water in their immediate vicinity. This structuring of water by components of cell membranes is thought to be of great importance. It will affect not only the actual packing tightness of the constituents of the membranes, but it may also alter the rate at which substances pass through the water layers attached near and in the membrane phase. All membranes have these structured water layers attached to their surfaces. They are called unstirred layers. This name is given to them because no matter how violently the solution bathing the cell is stirred, they appear to remain highly structured and keep closely fixed to the membrane's surface. They thus form another "membrane" of structured water and may themselves be a barrier to the free diffusion of small molecules. This aspect of water structure and its influence on membrane permeability is dealt with more fully in Chapter 8.

### The state of the membrane – a summary

We are now at the end of the first part of the book. This section has reviewed the experimental evidence about the structure of the cell membrane. We should look back at the progress we have made since Robert Hooke described his "cells" in 1665, some 300 years ago. A partial summary of the five early basic ideas of the cell membrane has already been made on page 21. It is interesting to see how they have to be modified in the light of more recent studies.

Without doubt the simple Danielli–Davson paucimolecular model of the cell membrane cannot now explain all the experimentally observed structures laid bare by positive and negative staining and electron microscopy. Certainly, as a rough and ready approximation, cell membranes can act as if they were bilipoidal barriers with "functional" aqueous pores. Whether the pores are a real structure in the membrane is still a matter of conjecture.

The new concept that membranes consist of basic nesting units containing structural protein with lipid asymmetrically adsorbed on to the inner and outer surfaces now seems to be a real possibility. Loosely nested, they would allow water and small hydrophilic molecules through, tightly nested they would become much more like a lipid barrier and hold back penetration of non-lipid soluble molecules. It is even possible that the arrangement of lipid and protein may change in the membrane. Some lipids might rearrange themselves from a bilayer to a micelle or clump of lipid leaving a space for small molecules to diffuse through (Fig. 21). The micelle may then go back to the more stable bilayer. Such transient formations of "pores" may be a feature of the living as opposed to the dead membrane seen after fixing, staining and blasting a beam of electrons through!

# CHAPTER 5

# The Movement of Molecules Across Cell Membranes

The previous chapters developed the idea that the outer plasma membrane acts as a diffusion barrier between the external environment and the internal world of the cell. The advantages of such a barrier are clear. The delicate molecular structures of life are not only protected from a changing and possibly harmful external environment but they are also maintained at conditions optimal for their biological function. The other side of the coin, however, is that the cell membrane barrier will delay or prevent not only the inward passage of nutrients essential to the welfare of the cell but also the outward passage of waste products from metabolism. Furthermore, in cells that manufacture substances for export, such as hormones or enzymes, mechanisms must be present to allow passage of these without damaging the protective barrier function. Numerous cellular mechanisms have in fact developed to overcome these problems of transfer. It will be useful at this point to make a general survey of the transfer systems that occur across biological membranes. We can then examine each in greater detail and discuss its importance to cellular physiology.

Rather than list the many cellular transfer systems, an attempt has been made to classify them under various headings in Fig. 22. This has been done to avoid the devastatingly barbed comment of the French novelist, Albert Camus that "The absurd man multiplies what he cannot unify!" There is, however, the danger of absurdity in making a classification of transfer systems at our present state of knowledge, for some systems can be made to overlap into different categories. This is not a real problem as long as we remember that the classification is only for convenience. It can and should be altered as soon as further work makes such changes advisable.

## Macro- and microtransfer systems

Movement through membranes can be conveniently divided into macrotransfer and microtransfer systems. Macrotransfer is

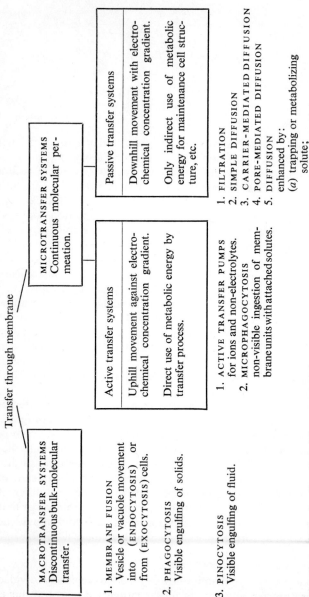

Transfer through membrane

**MACROTRANSFER SYSTEMS** Discontinuous bulk-molecular transfer.

**MICROTRANSFER SYSTEMS** Continuous molecular permeation.

Active transfer systems

Uphill movement against electro-chemical concentration gradient.

Direct use of metabolic energy by transfer process.

1. ACTIVE TRANSFER PUMPS for ions and non-electrolytes.
2. MICROPHAGOCYTOSIS non-visible ingestion of membrane units with attached solutes.

Passive transfer systems

Downhill movement with electro-chemical concentration gradient.

Only indirect use of metabolic energy for maintenance cell structure, etc.

1. FILTRATION
2. SIMPLE DIFFUSION
3. CARRIER-MEDIATED DIFFUSION
4. PORE-MEDIATED DIFFUSION
5. DIFFUSION enhanced by:
   (a) trapping or metabolizing solute;
   (b) pH gradients (NON-IONIC DIFFUSION).
6. CODIFFUSION

1. MEMBRANE FUSION Vesicle or vacuole movement into (ENDOCYTOSIS) or from (EXOCYTOSIS) cells.

2. PHAGOCYTOSIS Visible engulfing of solids.

3. PINOCYTOSIS Visible engulfing of fluid.

*Fig.* 22. A simple classification of the systems and types of mechanism that transfer substances across biological membranes.

movement at the bulk-molecular level and is discontinuous. It is the method by which cells move bulk solids or fluids into or out of the cytoplasm. Such transfer is of course highly dependent on a supply of metabolic energy. Examples of macrotransfer are found when cells engulf bacteria or when hormones or enzymes are secreted via vesicle bound granules. Microtransfer is transfer at the molecular level and is continuous. The systems usually handle individual molecules or ions passing to and fro although there are mechanisms that handle solutes in bulk. The distinction between this type of bulk transfer and the previous type is that in microtransfer the molecules in bulk are never aggregated, joined in a specific structure or surrounded by a vesicle or vacuole. While only a relatively few cells have macrotransfer systems all cells are thought to possess micro transfer systems for handling nutrients and ions.

### Active and passive transfer

A further useful distinction in microtransfer that we can make is to label the transferring system as either an active or a passive one. Active transfer was defined by Thomas Rosenberg in 1954 as transfer or movement against an electrochemical concentration gradient. This means that the substance has to be moved uphill against an unfavourable gradient of concentration and, if it is charged, against an unfavourable electrical potential difference. Some people have used the term "active transfer" when cells apply some of their metabolic energy to regulate and influence the rate and direction of transfer, but as this is not always easy to assess this definition is not as useful as the previous one. Passive transfer is said to occur when molecules or ions are transferred according to the prevailing electrochemical gradients, in other words when they move from a region of high to one of lower activity. It is possible that substances may pass through or utilize an active transfer mechanism during this downhill movement but we have no way of telling whether this is the case. A simple analogy is when a car is being driven downhill. A distant observer cannot tell whether the car is running downhill under the influence of gravity alone or whether the driver also has his foot lightly depressing the accelerator.

The simple distinction between active and passive transfer systems can sometimes lead to difficulties. It is possible, by manipulating experimental conditions carefully, to make a normally passive transferring system act like an active transferring one. We shall discuss this type of behaviour when we deal with the individual mechanisms.

### Specific macrotransfer systems

1. **Membrane fusion.** The term "cytosis" was coined to embrace all phenomena involving the formation of membrane limited vacuoles or vesicles. The prefixes endo- and exo- are used to indicate processes when cells ingest substances by a vacuole (endocytosis) and when cells eject substances by a vacuole (exocytosis). The

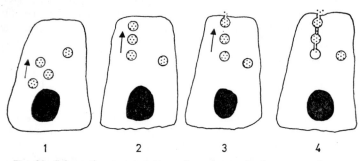

Fig. 23. Schematic representation of macrotransfer by exocytosis or cell secretion of granules. The process is like reversed pinocytosis and has been labelled emiocytosis (cell vomiting!). The intracellular granules or vacuoles surrounded by their membranous coat move up and this covering fuses with the outer cell membrane. This puts the granules or vacuolar contents outside the cell but the plasma membrane remains intact. In the above cell three vacuoles are involved in emiocytosis but only one remains in contact with the plasma membrane, the other two fuse with each other and the first. They pass their contents through each other to the outside.

vacuole may contain either fluid or solids while in other cases it may contain both. It is not known for certain how substances that are inside the membrane-enclosed vesicles leave the cell for the external environment or gain access from the vesicle to the cytoplasm. A number of cells appear to use the process of fusion of the vacuole or vesicle membrane with their plasma membranes. The contents of the vesicles are then either expelled from, or injected into the cell. By this mechanism the cell keeps the membrane barrier intact and prevents ingress or egress of other substances. In some cells a chain of intracellular vesicles may link up with one another by membrane fusion, the last one fusing its membrane with that of the plasma membrane. The contents of the vesicles are then passed through the chain until all are empty (Fig. 23). This type of macro-transfer is seen in pancreatic cells when their zymogen granules discharge their enzymes into the pancreatic juice.

An unusual and unique bulk transfer of material, perhaps the most important bulk transfer process in biology, begins when the spermatozoa crashes into the outer covering of the ovum. Gradually the sperm head is passed into the ovum so that its coded package of male DNA can link up with that of the female DNA and begin stamping out the proteins for the formation of a new living unit. It is thought that membrane fusion also plays a role in this macro-transfer process.

### Phagocytosis

This word, derived from the Greek, means cell eating. It was the term used by Elie Metchnikoff in 1884 when he saw leucocytes ingesting pigment particles and bacteria. These cell-eating cells (phagocytes), he surmized, must represent one of the basic defences of the human body against bacterial invasion. He received the Nobel prize for his studies in 1908. Phagocytosis is then, the visible engulfing of solids by the cell flowing around the object (Fig. 24). How much extracellular fluid is trapped in the engulfing process is difficult to tell. Once the solid is ingested and engulfed by a membranous vacuole it will be digested by intracellular enzymes if organic, or stored in the cytoplasm if inorganic and inert. Certain cells will phagocytose all manner of inert solids ranging from the carbon black particles of Indian ink up to small spheres of plastic. What we do not know is whether large single molecules can be phagocytosed into cells. Certainly large proteins are able to enter cells but how they do this is uncertain. Perhaps they become attached to a surface site and then this site is pulled into the cell with the adsorbed protein attached. This novel idea of a micro-transfer process by ingestion of intact membrane units, has recently been advanced by Leo Gross. In this process there would be no visible signs of membrane ingestion. Although the term "micro-phagocytosis" is thus somewhat inaccurate the process will be described under that heading in the relevant section.

### Pinocytosis

The word is also derived from the Greek and means cell drinking. It describes the process by which some cells visibly ingest the external fluid. G. Edwards appears to have been the first person to

describe cell drinking in amoeba in 1925 but the word pinocytosis
was first used by Warren Lewis in 1931, when he was observing
vacuole formation in macrophages and other tissue culture cells.
Since then it has been observed in many types of cells. The early

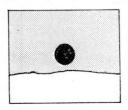

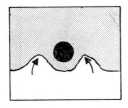

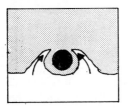

*Fig.* 24. Phagocytosis (cell eating). The cell membrane surrounds
a particulate object in the extracellular fluid. A complete membrane around the object is formed
by fusion of the two arms of the
projecting membrane. It is then
withdrawn into the interior of the
cell. Enzymes are secreted into the
vacuole to break down organic
material or the particle, if inert, is
stored in the cytoplasm. It is
thought (as the diagram shows)
that some of the extracellular fluid
is trapped with the particle when it
is engulfed. Thus the cell is probably pinocytosing (cell drinking)
at the same time.

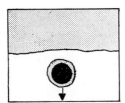

comparisons of phagocytosis and pinocytosis made much of the
possible differences between the two processes. Nowadays we note
how similar they are and regard them as different aspects of the
process of "membrane flow" (Fig. 25). A link between the uptake
of inert particles and solutes dissolved in the extracellular fluid has

been demonstrated in leucocytes. When these cells are made to phagocytose inert particles various solutes in the surrounding solution appear to enter the cell cytoplasm at the same time.

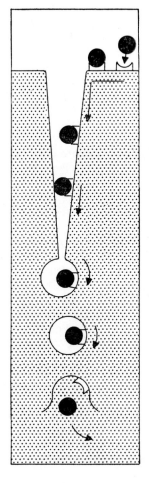

*Fig.* 25. The flowing membrane concept. Specific areas on the membrane are binding sites for various solutes in the bathing fluid. The membrane is in continuous movement, flowing into the cytoplasm and being pinched off to form small vesicles that are then broken down to release their attached solutes. This type of mechanism would give a high degree of selectivity to active transport by pinocytosis (cell drinking). The idea was put forward by H. Bennett in 1956.

Normally, without phagocytosis taking place, these solutes cannot enter the cell through the plasma membrane. This uptake of a combination of particulate and dissolved solids by inducing phagocytosis has been playfully called "piggy-back" phagocytosis.

In amoeba and other protozoa, pinocytosis consists of the surface membrane invaginating into the cytoplasm. The invagination is then sealed, often at the surface, and a membrane limited vacuole is formed which contains engulfed solutes dissolved in fluid (Fig. 26a). In leucocytes and tumour cells, pinocytosis is characterized by the plasma membrane making undulating and ruffling movements. Some of these lead to folds being produced which trap or enclose the outer fluid into vesicles. These are then withdrawn into the cell cytoplasm (Fig. 26b).

The vacuoles in both types of pinocytosis move into the interior of the cell. The force of movement may be great enough to break a thin mitochondrion in two! Whether the substances trapped inside the vacuole are liberated into the cytoplasm by the vacuole walls breaking down or by changes in their permeability is not clear. Visual evidence indicates the walls of vacuoles are far from rigid as they have been observed to shrink, coalesce, divide and be deformed. No one, however, has ever described a vacuole disappearing after it has been formed!

A number of compounds such as proteins and salts, when present in the external bathing medium, induce cells to pinocytose. The initial event that makes the external membrane invaginate or ruffle is thought to be the adsorption of the inducer molecule on to a particular site which then triggers off an energy releasing reaction that causes membrane movement. Although pinocytosis will allow cells to actively transfer substances either into their cytoplasm or from one compartment to another, it is unlikely to be a mechanism for the specific active transfer of nutrients. As the cell engulfs the external fluid it will indiscriminately absorb all the solutes present. It is possible, however, that some degree of specificity may be conferred if certain molecules from the external fluid become adsorbed to specific sites on the membrane before invagination or ruffling takes place (Fig. 25).

When an amoeba is pinocytosing really well the turnover of membrane is enormous, within five minutes 100 per cent of the membrane can be consumed and replaced! Metabolic energy must be used at a considerable rate for such a process. Studies with the electron microscope in 1953–54 by George Palade extended the concept of pinocytosis. He showed that microvesicles were present in a number of cells and suggested that they came about by pinocytosis at the submicroscopic level. This has now been termed "micropinocytosis".

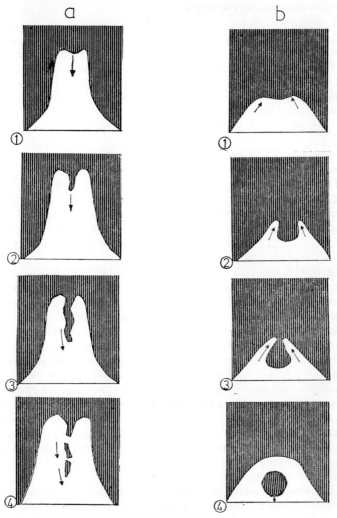

*Fig.* 26. Types of pinocytosis (cell drinking) observed in different cells. The first type (a, left-hand side) has been observed in amoeba and other protozoa. In these cells the plasma membrane indents itself and a channel is formed that breaks up into vacuoles. In the second type (b, right-hand side) observed in leucocytes and tumour cells, the membrane makes undulating movements and then folds over or engulfs the outside fluid entrapping it into a vacuole. In both types of pinocytosis the vacuoles move to the inside of the cells.

**Specific microtransfer systems**

1. **Filtration.** Substances can cross membranes by filtration. When a hydrostatic or osmotic pressure difference exists across a membrane water will flow in bulk through the membrane. The hydrodynamic flow can drag with it small molecules especially if they are significantly smaller than any holes, pores or discontinuities present in the membrane. This effect of water movement pulling solute molecules is called solvent drag.

Filtration occurs across a number of biological membranes. The primary formation of tubular fluid in the kidney is brought about by filtration of blood through the glomerular membrane. The water flow, filtering through the porous membrane, is accompanied by all the solutes of plasma except the proteins. The pore size is such that the membrane retains everything over the size of haemoglobin (molecular weight about 68,000). Filtration also occurs across the capillary wall "membrane" leading to the formation of a protein-poor tissue fluid. In both cases the energy for the filtration comes from the work done by the heart.

2. **Simple diffusion.** With simple diffusion the solute "dissolves" into the membrane phase on the outside of the cell, passes across the membrane matrix and then leaves the membrane for the aqueous, intracellular phase. The only driving force moving the molecule is thermal agitation and the concentration gradient. This type of movement through the membrane can clearly only occur with substances that are lipid soluble. Even with these substances, as Danielli pointed out some twenty years ago (page 19), once they have dissolved into the lipoidal membrane a large amount of energy is needed to cause them to break away and enter the intracellular aqueous environment. It is unlikely that many nutrient or hydrophilic substances enter cells by this mechanism but the gases oxygen and carbon dioxide, some compounds formed by metabolism and numerous foreign substances and drugs are thought to move across cell membranes by such a process.

3. **Carrier mediated diffusion.** In this mechanism, solute molecules attach themselves to a mobile "carrier molecule" which when loaded, traverses the lipoid membrane phase and releases the bound solute into the cytoplasm when it reaches the inside aspect of the cell membrane. The empty carrier is then thought to move back to the exterior of the cell by thermal agitation (Fig. 27). This type of "ferry boat" mechanism will allow lipid insoluble (hydrophilic) substances to traverse the lipid layers of the membrane. It is generally

called "carrier mediated transfer" and many substances of physiological importance are thought to be moved across membranes by
such processes. In the simplest type of mediated transfer the driving
process for the movement of the solute molecule is still thermal
agitation or the concentration gradient.

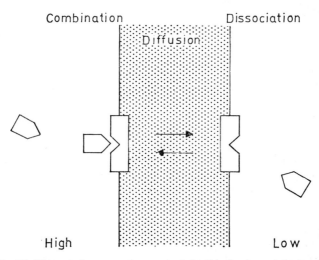

*Fig.* 27. The membrane carrier concept. In this fundamental concept of
transfer across biological membranes, a moving carrier or ferryboat molecule
ferries adsorbed solute through a lipid membrane. The process consists of three
parts, combination of solute with carrier, diffusion of carrier-solute complex
(which is more soluble in lipid than either alone) across the membrane to the
other face, dissociation of solute from carrier. If the solute is going from a high
concentration to a low concentration the energy of the concentration gradient
will power the movement of the carrier back and forth. The empty carrier will
move back across the membrane by thermal diffusion and pick up yet another
solute molecule.

4. **Pore mediated diffusion.** Up until now we have always assumed
that the molecule passing across the membrane moves through the
actual matrix or substance of the membrane. If, however, the membrane contains continuous aqueous channels or pores right through
this matrix, a small hydrophilic molecule could diffuse through the
membrane without having to leave the aqueous phase. The postulation that aqueous pores are present in cell membranes is nearly as
old as the concept of the membrane itself. Whether or not such pores

exist depends on the degree of confidence that is extended to the experiments purporting to prove their existence. Some cell physiologists passionately believe in pores (on the admittedly strong evidence available) while others reserve their judgement. There is little doubt, however, that pores would represent an important route for the movement of small molecules such as ions and water. Before we leave pores, one other mechanism of transfer that depends on such structures should be mentioned even though it is not strictly pore mediated diffusion. This is the movement of water that occurs when an electrical potential is imposed across a porous membrane. Water, being a charged substance (due to ionization and/or adsorption of ions on to the membrane), moves to the electrode which carries the charge of opposite sign. This movement has been termed electroosmosis. At present it is not thought very important in normal biological transferring processes.

5. **Solute diffusion enhanced by trapping, binding or metabolic transformation.** In this situation the solute moves across the membrane by any of the previously detailed mechanisms. Once inside the cell the solute is then removed or sequestered so as to effectively lower its intracellular concentration. It can be partially or completely removed from the intracellular aqueous phase by binding to proteins, trapping in a vacuole or even by metabolic transformation to other compounds. All these trapping mechanisms, because they have the common factor of lowering the intracellular concentration, will enhance the movement of the substance into the cell by creating an increased extracellular to intracellular concentration gradient. The diffusive movement of the solute is thus faster than that of its simple unaided diffusion or even its carrier mediated diffusion.

6. **Solute diffusion enhanced by pH gradients (non-ionic diffusion).** This case of transfer across a membrane is fairly complicated and rests on two basic features. The first is that many substances exist mainly as an ionized molecule or as an unionized, neutral molecule depending on the pH of the solution they are dissolved in. Such ionized substances are usually weak acids or bases. When ionized they are poorly soluble in lipid, but when in the non-ionized, neutral form they are often fairly soluble in lipid. This allows a transfer process to be generated across a lipid membrane simply by changes in pH across the membrane. Thus if we alter the pH on one side of a membrane to a low value (i.e. make it acid) we will make on that side, the ionization of weak bases easy but hinder the ionization of weak acids. The non-ionic neutral weak base will dissolve in the

membrane, diffuse across and become fully ionized at the low pH. This effectively traps the base on that side as the ionized molecule will not pass across the membrane (Fig. 28). Similarly if we make the

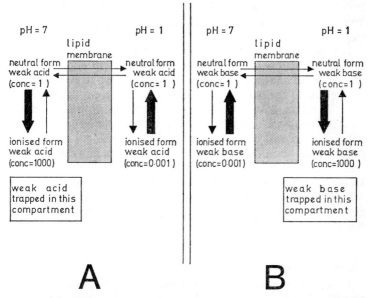

*Fig.* 28. Non-ionic diffusion – the influence of pH on the distribution of weak acids and bases between two compartments separated by a lipid membrane. In A, the weak acid present in both compartments will move from the compartment of pH 1 to that of pH 7. This is because the neutral molecule or non-ionic molecule is more soluble in the membrane than the ionized form. At pH 1, most of the acid will be in the unionized or neutral form (thick black arrow). This neutral form diffuses across the membrane to the pH 7 compartment where it then ionizes (thick black arrow). The ionized form of the weak acid now cannot diffuse back across the membrane. Hence the weak acid becomes accumulated in the pH 7 compartment. In B a similar argument applies to a weak base except in this situation the base becomes accumulated in the compartment of low pH, i.e. pH 1. Although the concentration of the neutral form may be the same and at equilibrium on each side, the total concentration of acid or base on each side will be very different.

pH on one side of the membrane correspondingly high (i.e. increase its alkalinity) weak acids will dissolve in the membrane in the neutral or non-ionic form. Then, when entering the high pH compartment they will ionize completely and are thus prevented from returning through the membrane (Fig. 28). They will be trapped at high concentrations on the high pH side of the membrane. If the metabolism of the cell is used to maintain the pH of the compartment into which the weak acid or base is diffusing, a continuous trapping process will be in operation that will effectively appear to concentrate the weak acid or base, just like an active transfer process.

7. **Codiffusion.** This type of movement is an example of the frictional interaction between molecules or ions. In codiffusion the movement of one ionic or molecular species down its concentration gradient aids the movement of a second, different species in the same direction. Like solvent drag, where the moving fluid entrains solutes (water–solute interaction), moving solutes by colliding with other solutes will entrain them in their directional stream. Codiffusion can be of two types. Either a solute movement causes another solute to move with it (solute–solute interaction) or a solute movement causes water to move with it (solute–water interaction). We do not know yet how important all these various interactions are in the transfer of substances across cell membranes. Recent studies suggest that solute–water codiffusion is a possible way of transferring water across membranes.

### Specific active transfer systems

1. **Active transfer pumps for ions and non-electrolytes.** Active transfer pumps are really only a special case of carrier mediated transfer. The one important difference is that all active transfer mechanisms must be able to concentrate solutes against their electrochemical gradients. Clearly for this type of activity, work has to be done on the solute transferred, the actual pump directly consuming the energy. Somehow the energy derived from metabolism is coupled to drive the carrier back and forth across the membrane. Pumping of the transferred solute will go on continuously as long as fuel and oxygen are supplied to the cell. Few active transfer mechanisms can function anaerobically.

For non-electrolytes, active transfer occurs when the substance is moved uphill against its concentration gradient. When we come to substances that have a charge on them like ions, amino acids or fatty acids, we have to be sure that they are not only pumped against their concentration gradient but also against their electrochemical gradient. Obviously if a membrane is polarized so that one side is negative and the other positive, ions and other charged substances with a net positive charge will move through the membrane and collect on the negatively charged side and vice versa. The charged substances may accumulate to a much higher concentration than that on the other side. There is thus a danger of saying that these substances are actively transferred by biological pumps. Our definition of active movement as transfer against the electrochemical concentration gradient excludes such a mechanism.

Two other processes can bring about transfer of non-electrolytes against a concentration gradient without there being genuine active transfer. If a carrier mediates the transfer of two substances with different affinities for the carrier the interaction between their movement on the carrier can cause one of the substances to be transferred against its concentration gradient. This is especially so if the other substance is moving down its concentration gradient, via the carrier. The energy from this downhill movement can power the uphill movement of the other. This process is known as "induced counterflow". It is also known that the active transfer of one solute can induce the apparent active transfer of another (primary active transfer inducing secondary active transfer). This is thought to happen in the mucosal cells of the small intestine. The pumping of sodium out of the cell by active transfer is thought to be intimately linked with the cell's ability to concentrate glucose and other sugars. Both induced counterflow and the ability of a primary active transfer mechanism to generate secondary active transfer of other solutes will be discussed in detail in the next chapter.

Because of the difficulty of defining active transfer in absolutely specific terms many cell physiologists would prefer to see the abandonment of the term. It causes more trouble, they argue, than it is worth. Hans Ussing, a leading figure in the field of transfer across biological membranes, has tartly defined active transfer as that transfer which cannot be explained by physical forces, e.g. electrical forces, diffusion and solvent drag. It may well be that in the future we can replace the term by the specific types of molecular mechanisms that transfer solutes. At the present it still remains a very useful general term for a particular class of accumulating mechanisms so that it need not be discarded on the junk pile of outdated terminology.

**Microphagocytosis.** Although phagocytosis is a well established macrotransfer mechanism by which cells ingest solids by forming a membrane vesicle around the object, little or no attention has been given to possible mechanisms by which small amounts of membrane are pulled into the cell interior without any vesiculation. If the plasma membrane was composed of small sub-units linked by calcium ions, which is a distinct possibility, these sub-units could be removed from the main membrane either by being replaced by another unit or perhaps by the other units expanding and sealing the gap across (Figs. 29 and 30). This novel idea of the cell ingesting its own membrane sub-units was proposed by Leo Gross in 1967.

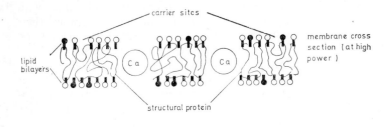

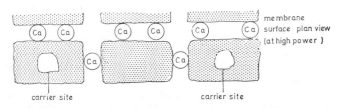

*Fig.* 29. A simplified diagram of the unit repeating membrane. The top side view consists of the units made up of a protein substructure upon which are hung the lipid bilayers. Some of the units have special carrier sites on their outer face that bind specific solutes in the bathing fluid. The units are probably linked together to form a sheet membrane by calcium ions (after Gross 1967).

If the sub-units were the parts of the membrane with the specific binding sites for amino acids, sugars or ions it would be possible to actively transfer these solutes into the cell. The tearing of the sub-unit from the membrane not only needs metabolic energy but also a store of preformed sub-units for replacement. The scheme for microphagocytotic transfer is shown in Fig. 30. It is of course both highly speculative and provocative. Careful thought needs to be given as to whether it could be energetically possible. Tearing the sub-units from the main membrane and replacing them may use more energy than can be provided by cell metabolism.

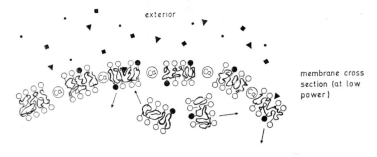

*Fig.* 30. How the repeating unit membrane can undertake **active** transfer. Each repeating unit is attached to the others by calcium ions. When a site is filled with a specific solute from the bathing solution this activates a change that allows the repeating unit to split off from the membrane proper. A new presynthesized unit takes its place thus sealing the gap. The energy for the disruption of the membrane comes from ATP hydrolysis. When the unit with attached solute molecule enters the cell the solute diffuses off into the cytoplasm and the unit is either cannibalized for construction of further units or is pushed back intact into the membrane at another point.

## Molecular models for active transfer pumps

Many ingenious molecular mechanisms have been proposed to account for the ability of cells to pump hydrophilic substances up-hill through lipid membranes. Although each model has its advantages and disadvantages, it has not yet been possible to devise critical experiments to find out whether any of them really do exist in cell membranes. Even so, they have an important function in that they allow us to visually interpret biological pumps at the molecular level. But, as they are only models, they should always be treated with a great deal of caution. Some of the most popular ones are shown in Figs. 31, 32 and 33. Most are self-explanatory. They all depend on a moving binding site or membrane. This gives specificity and the ability to pump. In one a contractile protein with its binding site on the outside is induced to fold up and contract so that its binding site with attached molecule moves to the inside of the membrane (Fig. 31). A second has a specific binding site on a swinging or rotating molecule (Fig. 32). The site binds the solute on the outside. Energy then moves the carrier around to the inside

Outside                                           Inside

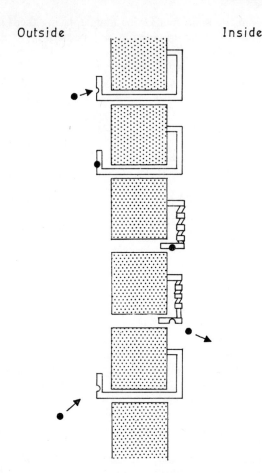

*Fig.* 31. A model for active transfer – the contractile protein. This speculative idea for active transfer uses a carrier site on a long chain protein molecule. The site is exposed to the fluid bathing the outside of the cell membrane. If a solute (black circle) becomes attached to the site this triggers off a reaction inside the cell whereby the protein is made to coil up, pulling the solute through the membrane into the intracellular fluid. The coiling process makes the site become less attractive to the bound solute which is released into the intra-cellular fluid. The protein then uncoils and again exposes its site through the membrane to pick up another solute molecule. The energy for contraction could come from ATP hydrolysis. The mechanism, by working in reverse would secrete substances from the cell (Mechanism based on Goldacre 1952).

where the solute molecule is released. The empty carrier then swings back to the outside to pick up another solute molecule. A third type of pump uses a membrane that goes from an open state to a collapsed, closed state. When open, solutes move in and bind to

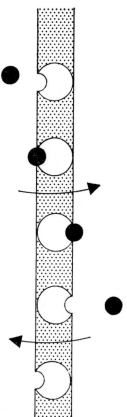

*Fig.* 32. A model for active transfer–the rotating carrier. This model utilizes a carrier site attached to a molecule that can rotate in the membrane matrix. It picks up the solute on the outside of the cell, then solute – carrier complex swings round or is rotated by application of an energy source and the solute is released inside the cell. A reverse process would give cell secretion of any constituent inside the cell.

various sites, then the membrane collapses asymmetrically and causes the substances to be released into the cell (Fig. 33). A final mechanism would be that of microphagocytosis as previously illustrated in Figs. 29 and 30.

**The forces involved in movement across membranes and their relation to systems of transfer.** Although many mechanisms have been suggested for the transfer of substances through membranes, the known physical and chemical forces that make such transfer possible are fairly limited. It may seem at first pointless to distinguish between mechanisms of transfer and the forces involved.

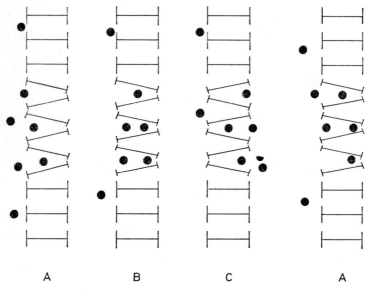

A                    B                    C                    A

*Fig.* 33. A model for active transfer – the asymmetric collapsing membrane. In this model for active transfer of solutes the membrane is open to the extracellular environment but closed to the intracellular environment in A. Solutes and fluid diffuse into the open matrix as in B. Specificity can be accomplished if the matrix has special binding sites for particular solutes. The membrane is then collapsed asymmetrically by application of energy (hydrolysis of ATP?) and the solutes are exposed to the inside of the cell. They then diffuse off the sites or away from the membrane matrix and the membrane again closes asymmetrically and becomes exposed to the outside solution.

Perhaps it is easier to comprehend the importance of the distinction if we use another motor-car analogy. Superficially there appear to be all manner of methods of powering cars. Steam, internal combustion, gas turbine and electric motors have all been employed as motive power. Even the methods by which the power is coupled to the driving wheels are varied. Yet, in the final analysis, the fundamental forces powering different cars are usually chemical or electrical. In biology we do not always know all the forces involved in powering movement but it is fairly obvious that the major ones will be (1) electrical, (2) chemical, (3) osmotic, and (4) hydro-dynamic. It is the task of basic research to discover how these forces operate in each specific transfer mechanism. Clearly in one as complicated as phagocytosis, many of the forces are likely to be operating at the same time. With simple filtration, however, hydrodynamic forces could well explain most of the transfer.

# CHAPTER 6

# Experimental Methods and Criteria for Characterizing Membrane Transfer

The previous chapter summarized the types of transfer that can be expected to take place across biological membranes. It is often easy to separate theoretically specific mechanisms and to talk about them at length. The real difficulty comes when we enter a laboratory and have to undertake critical experiments to prove or disprove their existence and the role they play in the transfer of nutrients, ions and water. In this chapter we will examine in some detail the experimental criteria that can be applied to measurements of membrane transfer in order to ascertain how a particular solute is transferred. Obviously some transfer processes are much easier to measure and assess than others. It needs little experimental technique or equipment (nowadays at least) to see whether a cell can transfer fluid by pinocytosis or solids by phagocytosis. A good microscope, a slide and cover slip and acute observation are all that are needed. When it comes to proving that pores exist or that a compound prefers passing through pores to crossing via a carrier, both the techniques and the theory utilized are often extremely complex. Experiments in fact become so complicated that they often leave a small gap of uncertainty about the conclusion to be drawn. Biological experimentation at this level is extremely difficult because the experimenter never knows for certain if he has controlled all known variables except the one he wants to measure. Apart from the difficulty he has in holding constant the known variables there are unknown ones impossible to control. This is why tomorrow's experiments often make today's conclusions inaccurate. Before we discuss the intimate details of the permeation of substances through membranes it would be useful to examine our ideas about the diffusion of molecules in more simple media like liquids.

## Diffusion in fluids

Because diffusion means random movement due to thermal agitation, it will always cause molecules or ions concentrated in a

particular volume of fluid to move out and try and make the surrounding fluid of equal concentration. Like death and the grave,
diffusion is a great leveller! As all cells obtain their oxygen and their
nutrients by diffusion from the circulating blood it is of interest to
calculate how efficient diffusion is as a transfer process. Or, to put
it in another way, how long would it take for a solute to travel into
a pure solvent and cause the concentration to increase up to 95 per

TABLE 9. *The speed of diffusion of a small molecule (nitrogen)*

| Distance over which diffusion takes place | Approximate time taken to reach 95% of the initial concentration |
|---|---|
| $1 \mu (10^{-4}$ cm) | 0·06 seconds |
| $3 \mu$ | 0·6  seconds |
| $10 \mu$ | 6   seconds |
| $100 \mu$ | 11  minutes |
| $1,000 \mu (0·1$ cm) | 18  hours |

cent of its original concentration? The data in Table 9 give some
idea of the time involved when a solute diffuses across various
distances. When these distances are small, of the order of the dimensions of cells, diffusion can achieve practically complete equilibrium of concentration within less than a second. As soon as the
distances become 1 mm. or more the time taken begins to be
measured in days. Thus, over small distances diffusion is highly
efficient, over long ones it becomes grossly inefficient.

**Diffusion kinetics**

1. **Fick's Law of Diffusion.** The process of diffusion was studied
by Fick in 1885. He developed a simple mathematical treatment on
the intuitive assumption that the flow of solute through a solution
was similar to the flow of heat along a bar. Fick substituted the
quantity of solute for the quantity of heat and the concentration of
solute for temperature. The rate of diffusion $(dn/dt)$, or the amount
crossing a given area A in an infinitesimally short interval of time,
should be proportional to the cross sectional area and to the concentration gradient $(dc/dx)$. The movement of a substance through
a liquid will also depend on the number of impacts that the solvent
molecules make on the diffusing substance and on the size of the
molecule. Large, diffusing molecules will be impeded by the solvent
more than small ones. The viscosity of the solvent will also affect

the diffusion. All these interactions and variables can be compounded together to give a number called the Diffusion coefficient, D. Its dimensions are cm²/sec, and it becomes smaller as the viscosity of the solvent increases or as the size of the diffusing solute increases. We can now relate the rate of diffusion to the various factors in the following form:

$$\text{rate of diffusion (dn/dt)} = -D.A. \, dc/dx$$

The negative sign shows that the solute is diffusing from a region of higher concentration into a lower one. The equation is known as Fick's Law of Diffusion. It allows us to make some simple predictions about the rate of diffusion. For example, the greater the concentration gradient $dc/dx$, the faster the rate of diffusion. Furthermore this rate must be directly proportional to the concentration gradient. Thus, if simple diffusion is in operation across a biological membrane, increasing the amount of substances on the side from which the solute diffuses should cause a corresponding increase in the amount transferred. There will be a simple linear relationship between amount transferred and concentration and no saturation will occur provided we use sensible concentrations of substances (Fig. 34). If excessively high concentrations are used they could cause osmotic damage to the cells or change the structure of their membranes.

This application of Fick's Law of Diffusion to biological membranes has been of great importance. It allows a simple experimental approach to determine whether the transfer of a substance across a biological membrane can be explained satisfactorily by the kinetics of simple diffusion or whether other processes have to be invoked. In many respects the application of so simple a law to such complex membranes really entails enormous oversimplification. Fortunately, the application has been eminently satisfactory as a first approximation. If we start to examine the process of membrane diffusion in a rigorous manner, taking into account interactions between membrane and solute, the solute and the membrane, water and the solute, we find ourselves involved in fearsome equations which can be solved only by a knowledge of higher mathematics. This is not only outside the scope of this book but also out of the scope of the author's own mathematics! We shall be content then, like many other biologists, to use the simpler, but less accurate, Fick's Law! There is, however, one change that we can undertake to make the equation more useful.

**The permeability constant.** In the original Fick equation for diffusion in fluids the concentration gradient is the differential term $dc/dx$. When used for practical membrane studies we can modify the equation to avoid this differential by assuming that the concentration gradient across the membrane is equal to the concentration outside the cell (Co) minus the concentration inside the cell

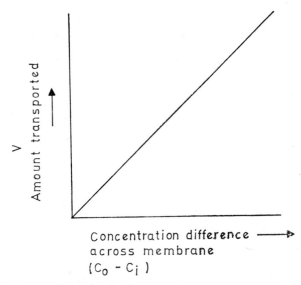

Fig. 34. The kinetics of movement of a substance crossing a membrane by simple diffusion. Increasing the concentration of the solute in the bathing fluids causes a direct increase in the amount of substance transferred across the membrane. The relationship between the concentration difference across the membrane (C out – C in) and the solute transferred is thus linear and follows Fick's law of diffusion.

(Ci) divided by the thickness of the membrane (d). The equation becomes,

$$\text{rate of movement} = -D.A. \frac{(Co - Ci)}{d}$$

This equation, however, now includes the thickness of the membrane as an unknown. The early part of the book was at pains to show how difficult it is to estimate the thickness of cell membranes with any accuracy. Because of this, it would be better to modify the equation once again so as to remove this term. We can also avoid having to measure the area of the membrane by making the three terms

$\dfrac{-DA}{d}$ into a new constant called the permeability constant. This

constant is defined as the number of moles of substance which cross a unit cross section area of membrane in unit time under unit concentration difference, it thus has the dimensions of a velocity, viz. cm/sec.

2. **The kinetics of carrier mediated diffusion.** In the kinetics of simple diffusion, increasing the solute concentration will cause

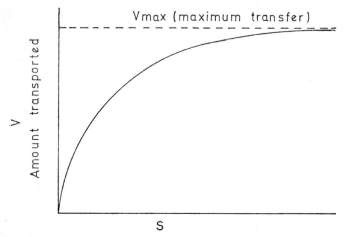

*Fig.* 35. The kinetics of movement of a substance crossing a membrane by carrier mediated transfer. The kinetic relation between solute concentration of the bathing fluid (S) and the amount of solute transported (V) is not linear, increasing amounts of solute are carried with increasing solute concentration until at a high level all the carriers become filled and the transport process becomes saturated. This maximum transfer capacity is known as the $V_{max}$ of the system.

increased transfer. Would the same effect occur if the solute was transferred through the membrane by a carrier? Are carrier kinetics different from those of simple diffusion?

Let us describe the process of carrier diffusion in words and then convert this description into a more useful mathematical one. First there is the association process. The solute approaches the membrane and unites with the carrier. The solute-carrier complex then moves across the membrane and when it arrives at the inside aspect of the membrane the solute is released into the cell. This is the dissociation process. The carrier then moves back by thermal agitation to the external membrane surface to pick up and ferry another solute molecule. Increasing the external concentration of solute in this system would at first cause an increase in the number of solute molecules attaching themselves to free carrier sites, thus an increase in transfer across the membrane would result. This would continue until all the carriers that were available were filled and were transferring solute across the membrane. At this stage all the transfer sites would be saturated and further increases in the transfer on increasing the solute concentration would not occur. The transfer in fact would level off or plateau at this saturation concentration (Fig. 35). The transfer at this saturation concentration is called the maximum transfer or "V max". Thus one very typical feature of a carrier mediated transfer system is that it will show "saturation kinetics".

While the above description of carrier mediated transfer is useful it does not allow us to quantitate or measure in any way the relations between the solute and the carrier. If we treat, however, the reaction between the carrier and the solute transferred just like the reaction between an enzyme and its substrate, we can obtain useful equations that allow us to calculate, without too much difficulty, the relations between carrier and solute.

**Carrier-solute reaction treated as an enzyme substrate reaction**

All reactions between enzymes and their substrates can be portrayed in a simple chemical equation of the type:

$$\begin{matrix} \text{free} \\ \text{enzyme} \end{matrix} + \text{substrate} \begin{matrix} \leftarrow \\ \rightarrow \end{matrix} \begin{matrix} \text{enzyme-substrate} \\ \text{complex} \end{matrix} \rightarrow \begin{matrix} \text{free} \\ \text{enzyme} \end{matrix} + \text{products}$$

The first part of the equation has the enzyme-substrate complex (ES) in equilibrium with the free enzyme (E) and substrate(S). The ease with which this complex dissociates into its constituents can be represented by a constant known as the dissociation constant and designated as $K_s$. Two biochemists, Michaelis and Menten, used the above equation to describe the interaction of an enzyme with its substrate and found that there was a definite mathematical relation between the dissociation constant of ES into E and S ($K_s$), the velocity of the chemical reaction between E and S (v), the substrate concentration (s) and the maximum velocity at which the reaction could take place (V max). The relation was

$$v = \frac{V \max.}{\left(1 + \dfrac{K_s}{s}\right)}$$

If the appropriate values are placed in this equation it will generate the curved graph seen in Fig. 35. Curves and equations of this type are not easy to handle mathematically. Michaelis and Menten found, however, that the equation greatly simplified if they chose a particular value for the substrate concentration "s". When they made "s" equal to $K_s$, that is when the solute concentration became equal to the numerical value of the dissociation constant of the enzyme-substrate complex, the equation simplified thus:

$$v = \frac{V\,max}{1 + 1} \quad \text{or} \quad v = \tfrac{1}{2} V\,max$$

This means that when the concentration of the substrate becomes equal to the dissociation constant of the enzyme-substrate complex, the velocity of the reaction the enzyme catalyses will be half the maximum velocity. This "dodge" allowed such a great simplification that the actual value of "s" needed to bring it about was given a special name, that of the Michaelis constant or Km (usually pronounced just as it is printed, K M). It is the concentration at which the enzyme is half saturated with substrate. This is in fact also a measure of the affinity that the enzyme has for the substrate or vice versa. A Km has the dimensions of concentration, viz. mM/1 or mg/100 ml. If an enzyme reacts with a number of different substrates then each substrate will have a different Michaelis constant or Km. Such a measure of the affinity an enzyme has for its substrate is obviously of great value in experimental studies. One difficulty still exists. How do we obtain the Km from our experimental data or from the awkward equation that describes our curve in Fig. 35. Two other scientists solved this problem with yet another ingenious arithmetical dodge. These research workers, called Lineweaver and Burk, performed a mathematical juggle that allowed the Km to be obtained from a simple graphical plot. They found that they could make a startling transformation of the saturation-kinetic curve of Fig. 35. Instead of directly plotting the velocity "v" against the substrate concentration "s" they simply plotted the reciprocals of "v" and "s", that is they plotted 1/v against 1/s. When this was done the curve was transformed into a straight line (Fig. 36). Mathematically speaking, straight lines are far easier to play with and handle than curves of any description. Furthermore, this straight line had some wonderful properties. If it was extrapolated back to cut the X-axis, the point at which the line cut the axis was found to be the negative reciprocal of the Km, i.e. −1/Km! At last one could easily measure the affinity of enzymes for various substrates and have a simple quantitative way of comparing these affinities. There is one difficulty, however, in the relation of the actual numerical value of a Km to the type of affinity an enzyme has for its substrate. A high value of the Km means a low affinity of the substrate for the enzyme and not, as would be intuitively expected, a high affinity. Low values of Km indicate a high affinity of the enzyme for the substrate.

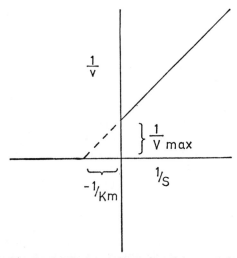

*Fig.* 36. The Lineweaver–Burk transformation of the saturation curve shown in Fig. 35. By plotting $1/S$ against $1/V$ the curve is transformed into a straight line. Where this line cuts the abscissa is in fact $1/V_{max}$ and where it cuts the ordinate (if extrapolated, dotted line) is $-1/K_m$. The Lineweaver–Burk plot is extensively used in enzyme-substrate systems and for carrier-solute systems. It allows one to estimate the affinity of the carrier for the solute, i.e. the $K_m$ or half-saturation constant of the carrier.

How can we use this newfound quantitative measure of the affinity of a substrate for its enzyme in relation to our carrier in the membrane? Well, if we make the assumption that carriers are like enzymes and that just as enzymes bind their substrates the carriers bind their solutes in a solute-carrier complex which then dissociates when it arrives at the inside of the membrane, we can hopefully use exactly the same formulation of carrier kinetics as was used for the enzyme-substrate reaction. We simply substitute for velocity of the chemical reaction the velocity of transfer of the solute across the membrane (i.e. rate of transfer) while the "substrate concentration" merely becomes the "solute concentration". The dissociation constant of the enzyme-substrate complex becomes the dissociation constant of the solute-carrier complex. It is still written as Km but it is the measure of the affinity a solute has for the transfer process. Now let's see what happens in a real experiment using this type of carrier kinetics.

### Sugar movement across red cell membranes by carrier mediated diffusion

The experiment entails the use of red blood cells. We incubate separate samples of these cells with increasing concentrations of a single sugar, say glucose, and measure the rate at which the sugar enters the cells either by chemical or physical methods. We can then plot a graph of rate of entry, "v", against the glucose concentration, "s", of the incubating fluid. We should obtain a curve like Fig. 35 revealing that the carriage of glucose into the cell shows saturation kinetics. As this curve is awkward to handle, it would be better to use the reciprocal plot of Lineweaver and Burk. By making a graph of 1/v against 1/s we should obtain the straight line of Fig. 36. Extrapolating the line to cut the X-axis we can obtain the value of the intercept; this is equal to $-1/Km$. We have thus obtained a Km value for the affinity of glucose for the red cell carrier mediated transfer process. Now we want to know whether the carrier of this transfer process can handle any other sugars. So we repeat the whole experiment but this time with a new sugar. Again we run a whole series of sugar concentrations as before and measure their entry at each level. We plot a new Lineweaver–Burk graph for these results, extrapolate the line back and obtain a Km for this sugar. Similarly, we can repeat the whole process for any number of sugars we think might be handled by the carrier system. We thus collect a large number of Km values for each of the different sugars. Remembering that those with a low Km indicate a high affinity for the carrier while those with high Km have a low affinity for the carrier, we can immediately see the specificity of the carrier system for different sugars. In fact we can go one stage further, we can correlate this specificity with the molecular structure of the various sugars and see whether there is any structure or group best fitted or suited for this particular carrier site. In this account we were dealing with a hypothetical experiment but the data in Table 10 were in fact obtained by just such a series of real experiments. Human red cells were used in this case and the movement of a whole series of sugars across the membrane by a carrier mediated system was measured. Only a swift glance at the table is needed to see that glucose, with the low Km, has a very high affinity for the transfer site while its optical isomer L-glucose has hardly any affinity for the transfer site (high Km). Thus a small change in molecular structure can create a very large change in affinity of a solute for a transfer process. This is one of the important points about carrier mediated systems – they usually have a high specificity.

TABLE 10. *Affinities of different sugars for the carrier mediated transfer system in human red blood cells*

| Sugar | Km | Affinity of carrier for sugar |
|---|---|---|
| D-glucose | 0·005 | Very high |
| D-galactose | 0·03 | High |
| D-xylose | 0·06 | Medium |
| L-galactose | >3·0 | Very low |
| L-glucose | >3·0 | Very low |

The type of experiment we have described can be done with any carrier mediated system, even when such systems are linked to active transfer pumps. The simple concept of Km has been of enormous value in cell physiology. For example if two sugars have a low Km (high affinity) for a carrier mediated transfer process then it would be expected that if they were both present in the fluid they would compete for the same transfer site. We can thus, to a certain extent, forecast whether or not one substance will strongly compete with another for a transfer process by knowing the value of its Km. If the molecule can be structurally modified we can make new compounds that will compete with normal nutrients for the various cell transferring processes. It is thus possible to design with some precision drugs that block specific activities in cells. This technique has enormous potential in cancer therapy when we try to kill cancer cells but leave normal ones intact.

**The simple made more complex but more accurate.** In applying the Michaelis–Menten equations to our carrier kinetics we have of course made some big assumptions and grossly oversimplified the true situation. We took into consideration only one association-dissociation of the solute with the carrier when in fact, as any membrane has two sides, there must be two. One association-dissociation takes place at the outside interface and another at the inside interface. It is not too difficult to see that we can apply the same kinetic arguments to the carrier-solute interactions at both faces as we applied to the single face. This gives us a rather more complicated equation for the carrier-solute interaction but it is more accurate and it predicts the behaviour of many carrier mediated systems found in cell membranes.

### Criteria for assessing whether a system is a carrier mediated one

In the previous experiments with sugars and red blood cells we assumed from the start that the sugars were carried across the membrane into the cell by a carrier mediated process. If the physiologist is dealing with a tissue or cells that no one else has used, or a substance that no one else has studied, how can he prove by laboratory experiments that the substance transferred is ferried across the membrane by a carrier? One method would be to make a list of the properties that carrier mediated transfer systems in different tissues and cells have been shown to share. If the transfer of his substance fits one or more of these, then our physiologist is obliged to think that there is a mechanism aiding the movement of the substance across the membrane. The more properties his transfer fits, the greater the likelihood that it is genuinely a carrier mediated one. The basic characteristics that carrier mediated transfer systems must possess are:

1. **Rate of movement.** The rate of movement of the substance into the cells, or across the epithelial membranes, is a good deal faster than predicted from either its oil–water solubility coefficient, or comparison with other molecules of approximately the same molecular size and lipid solubility.

2. **Saturation.** The rate of penetration or transfer gradually reaches a limit as the concentration of the substance is increased. The system thus shows "saturation kinetics" and the simple law of diffusion proposed by Fick is not obeyed.

3. **Competition.** The rate of penetration or transfer of the substance is specifically depressed when other substances of similar molecular structure are added. The system thus shows "competition" for the transferring site or carrier molecule.

4. **Specificity.** Optical isomers of the substance or substances of similar structure are transferred at different rates than the test substance. The system thus shows a high degree of "specificity".

5. **Inhibition (of carrier site).** The rate of penetration can be depressed by various inhibitor substances that have a chemical structure quite dissimilar to that of the penetrating solute and that act at low concentrations. Often the inhibitor molecules are so large that it is unlikely that they can themselves penetrate through the cell membrane. Most of the inhibitors in use are enzyme poisons. They can be reagents that attack specific chemical groups, or they can be the ions of heavy metals such as copper, mercury, lead or uranium. Usually the amount of inhibitor needed to markedly reduce the transfer of the penetrating solute is only large enough to cover a small fraction of the surface of the cell membrane. The inference is that they attack or bind with the specific sites on the membrane that mediate the transfer of the solute, rather than penetrate the cells and interfere with metabolic pathways.

Bearing this list of properties in mind, our physiologist would thus be wise to demonstrate at least saturation kinetics, specificity and competition before he announces to the world that the new transferring system he has investigated is genuinely a carrier mediated one. If he did find that his system fulfilled all the criteria, then he would add yet another carrier mediated system to an already swollen list. The transfer of many nutrients (especially amino acids and sugars) and many ions occurs by carrier mediated systems in cells of bacteria, yeasts and the vertebrate and invertebrate phyla.

#### Induced counterflow

Although carrier mediated transfer is basically a mechanism for passive movement of solutes down their electrochemical gradients it can, under unusual circumstances, be made to pump a solute uphill. It does this by using the energy of another solute that is moving downhill. We can understand this process more easily if we make use of a model of a mobile carrier in a membrane that has an affinity for two solutes called solute H and solute L. On one side only of the membrane there are a number of H molecules while on both sides of the membrane there are equal concentrations of L molecules. The carrier has a greater affinity to link with solute H than with solute L. Using our new found measure of the affinity of carrier for solute, the $K_m$ of H for the carrier is smaller than the $K_m$ of L for the carrier. Now what happens when we start the system in motion? Both solutes H and L will collide with the outer membrane and its carrier, but as H has a greater affinity for the site than L, it will compete with L and tend to fill the site more frequently than L. Thus H will move across the membrane with the carrier until it gets to the other side where there is no H but a large concentration of L. H will leave the carrier and L will take its place. As H now has an infinite reservoir to move into, it is very unlikely that it will attach itself to the carrier on any subsequent journeys, moreover the chances of attaching in the presence of so many L's are low. Thus the carrier will make its journey back loaded with an L. When it arrives at the outside of the membrane the L will dissociate from the carrier because of the fierce competition for the site by the H molecules; the site will again be filled with an H and the carrier returns to the other side. What we have done then is to move an L molecule from one side of the membrane to the other. As both sides of the membrane had an equal concentration of L it is being transferred uphill against its concentration gradient! The more the carrier moves to and fro laden one way with H and returning with a cargo of L, the greater will be the transfer of L against its concentration gradient. Where is the energy coming from to keep this uphill transfer of L going? It certainly isn't coming from metabolism of the cell because in our model there is no metabolic apparatus to supply energy. If, however, the model is examined it will be noted that H is actually moving from a high concentration on the outer side of the membrane to a lower concentration on the inner side. Thus H movement down its concentration gradient induces a counterflow of L up its concentration gradient. When the concentration gradient of H is completely run down, so that it is equal on both sides of the membrane, the uphill pumping of L will stop. The process of induced counterflow is not true active transfer because the transfer mechanism does not directly utilize metabolic energy. There may be circumstances by which a cell initially builds up the high concentration of H by a biological process that uses metabolic energy. Under these conditions it might be extremely difficult to separate this use of metabolic energy from that directly feeding a pump.

### Criteria for assessing whether a transfer system is an active pump

As we have stressed previously the major criterion for ascribing that a substance is actively transferred is its uphill movement against its electrochemical concentration gradient by a pump that is directly using metabolic energy. Active transfer systems are thus sensitive to substances that depress cell metabolism, i.e. metabolic inhibitors. Unfortunately, most metabolic inhibitors have poor specificity so that they affect many cell functions other than the supply of energy to active pumps. Because of this, inhibition of active transfer by metabolic inhibitors cannot be used as a primary criterion for active transfer. Scientific folklore about metabolic inhibitors is legion. Someone once said that the degree of confidence that an experimenter had in the specificity of an inhibitor was in inverse proportion to his familiarity with its use! Another scientist joked that investigation of the metabolic pathways of a cell by using metabolic inhibitors was like dropping a bomb on a power station to see how an electric light bulb worked! Thus, the often quoted proof for active transfer usually found in elementary text-books – that it depends on metabolic energy – is a very poor criterion to choose by itself. If, however, it is added to the other criteria it gives another piece of evidence for such a process. Because active transfer pumps are the "special case" of carrier mediated transfer, all such pumps must have the essential re-quisites of a carrier mediated transfer system. Thus an active transfer pump must exhibit:

1. Movement against an electrochemical concentration gradient.
2. Saturation kinetics.
3. Specificity.
4. Competition.
5. Site inhibition.
6. Metabolic inhibition.

Even when we have this list of criteria for reference it is still not an easy task to make absolutely sure that a system that accumulates solutes is one that can be labelled an "active transfer". Two examples of specific experiments will show some of the difficulties.

### Problems of active transfer

1. **Intracellular activity.** Let us assume that we want to demon-strate that cells from a certain organ actively transfer substance A into their cytoplasm. We will make substance A a neutral, non-electrolyte so that the complications of transfer involving electrical potentials and non-ionic diffusion processes are not present. First we would incubate the cells in a solution containing a known con-centration of A. Let us assume that this initial concentration is 100 mg./100 ml. After one hour's incubation the cells are removed (usually by centrifugation), washed to remove any adhering

incubation fluid containing A and then ground up so that their intracellular content of A can be estimated. If this concentration is significantly greater than 100 mg./100 ml. (measuring the concentration in the intracellular water) say 500 mg./ml., would it be a case of active transfer of A? Could we say that an active pump for A has been found?

At first the temptation is to say yes, the concentration inside the cell is higher than that outside so the substance must have been pumped into the interior across the cell membrane. This would be true only if, A, when inside the cell, is in a free state. All that we have measured is its chemical concentration. What we really need to know is its activity. By this we do not mean whether it reacts chemically with other substances but whether it acts as a free molecule in solution. In technical jargon, is it still "osmotically effective" inside the cell? If it is then we have certainly demonstrated active transfer; if it is not, then we may simply have been looking at one of the numerous passive transferring systems documented in Fig. 22 in Chapter 5. How do we in fact measure whether the molecule is osmotically active inside cells? This is in fact not an easy thing to do. It has only been accomplished for a very few substances and cells. We have two approaches. Either we can directly measure the activity of the solute inside the cell or we can use indirect evidence about its activity. Direct measure is the preferable one but it is the most difficult. One direct method of measuring activity inside the cell is to measure the mobility of the solute. If the mobility of A inside the cell is nearly the same as that outside in the bathing solution it is highly unlikely that A can be bound inside the cell. Direct measures of mobility can, however, only be measured in very large cells such as the giant axons of the squid and some large plant and algal cells. The mobility of the potassium ions inside squid axon fibres has been found to be only 1–2 per cent lower than that in the bathing sea water so it is unlikely to be bound. Indirect measures of intracellular activity use the movement of water as an index of osmotic activity. If a solute is being accumulated inside a cell in a free state it will cause water to move into the cell by osmosis until the internal and external osmotic pressures across the cell membrane are equal. Hence, cells accumulating substances should increase their volume by swelling slightly. This swelling due to movement of water has been observed in a number of cells that accumulate sugars or amino acids to high concentrations. Few other substances have been tested to see whether they are osmotically active inside cells. There is some danger then in automatically assuming that because a substance has a higher intracellular chemical concentration than extracellular it must be a candidate for active transfer. When cells can transfer substances right across their cytoplasm into another compartment (trans-epithelial transfer), problems of this nature are much easier to solve. The osmotic activity of the transferred solute in free solution is much easier to measure than when it is inside cells. Because of this, cells that actively transfer solutes from one solution to another (like cells in the stomach, intestine, kidney and salivary glands) are often preferred as experimental models for investigating active transfer processes.

**Problems of active transfer**

2. **Primary and secondary uphill transfer.** We have always assumed that different transfer systems are quite distinct entities and that they are not linked. This, however, is often not the case. In real life, transfer systems can become interdependent, one system driving another. When this occurs with an active pump for one solute linked up with a carrier mediated transfer system for another solute, we can have most unusual effects. The active transfer system will supply the motive power to allow the carrier mediated transfer system to transport its solute uphill. In other words the primary active transfer system (pump) is linked to cause a secondary active transfer. Some people have called this process co-transport. The usual mechanism by which the two systems are linked is by a common carrier that handles both solutes together. Such a common carrier is shown in Fig. 37. It has two specific sites. One site is for solute A and the other site is for a second species, solute B. The carrier moves across the membrane to the inside aspect only when both sites are filled. Because there are three molecules involved in this association (solute A, solute B and the carrier molecule) it is called a "ternary complex". Once across the membrane the ternary complex is exposed to the intracellular fluid. If the concentration of solute B is kept low in this fluid by means of an active pump at the other membrane, solute B will dissociate from the ternary complex and enter the intracellular fluid. It will then be removed from the cell by the continuous pumping action of the active pump. With the loss of solute B from the carrier, the other (solute A) will also dissociate from the carrier and enter the cell. The empty carrier then recrosses the membrane to its outside face and picks up

another solute B and solute A molecule. The loaded carrier again returns to the inside face of the membrane and the whole procedure of solute dissociation is repeated. Solute B is always pumped out of the cell by its specific active pump but gradually solute A is accumulated intracellularly to a higher concentration than is

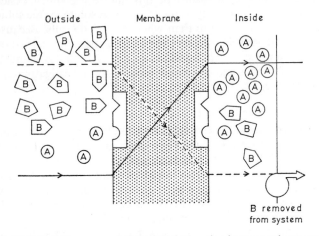

*Fig.* 37. Schematic representation of the interaction between solutes entering a cell via a common membrane carrier. The carrier has a site for solute A and solute B. If the carrier only moves across the membrane when it is filled both with A and B then both solutes will enter the cell. If B is continually removed from the cell (by metabolism or perhaps a pump) then A can be accumulated to a high intracellular concentration by using the energy of B going down its concentration gradient into the cell, via the linked carrier. Such mechanisms are well known and are called coupled transfer systems. Some are known to operate in intestine and kidney. The former type is shown in more detail in Fig. 38.

present outside the cell. The net effect of the operation of the solute B active pump and the linked carrier for solutes B and A is to drive solute A uphill, from the energy involved in solute B going down its concentration gradient into the cell on the joint carrier. In the last analysis it is clear that it is the metabolic energy expended on the active pump for solute B that is driving solute A uphill. In this model we have the unusual case of a primary active pump driving another uphill transfer system. Such an arrangement is not a hypothetical model dreamed up by a surrealistic physiologist but is thought to be the actual process by which the mucosal cells of the small intestine (and possibly the tubule cells of the kidney) carry on the transfer of sugars and amino acids. In the case of the small intestine, solute B is a sodium ion while solute A is a glucose mole-

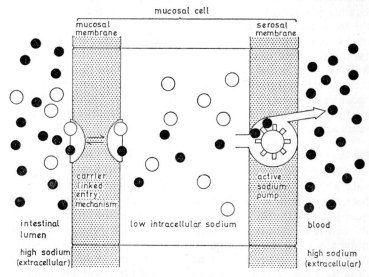

*Fig.* 38. A mucosal cell of the small intestine in diagrammatic form. The inside of the cell has a low sodium concentration which is maintained by the agency of the active sodium pump thought to be at the serosal membrane of the cell. This pumps the sodium from the cell into the blood stream. The mucosal membrane of the cell is exposed in the lumen of the gut to sodium (solid circles) and glucose molecules (open circles) when a meal is eaten. These two solutes attach themselves to the carrier which then moves across the membrane. The sodium ions (solid circles) leave the carrier on the inside of the membrane because the intracellular sodium concentration is low. Their loss from the carrier changes the affinity of the glucose binding site; it becomes less attractive to glucose which diffuses off into the cell. The carrier then moves back to the outside. If this process is constantly repeated the glucose enters the cell and is actively concentrated. The energy for its entry and accumulation comes from the sodium pump, the common carrier providing the link between sodium transfer and glucose accumulation.

cule (Fig. 38). The glucose molecule and sodium ion fit into their respective sites on the carrier forming the ternary complex (glucose–Na+–carrier molecule). This moves to the inside aspect of the membrane. The low level of intracellular sodium (kept this way by the active sodium ion pump at the far border of the cell) causes the sodium ion to dissociate from the ternary complex. When this happens the glucose molecule also becomes detached from the complex. The carrier then moves back to the outer face of the membrane to pick up and transfer more glucose and sodium. Gradually glucose is accumulated inside the cell and spills out into the blood and lymph and is carried away to be utilized by the rest of the body. Although the detailed description is for sugar transfer, exactly the same mechanism is thought to take place for the co-transfer of an amino acid and a sodium ion. One consequence of this unusual co-transfer is that the movement of non-electrolytes becomes intimately linked with the movement of sodium ions. This is a feature of biological transfer that was unthought of not so many years ago.

### On the nature, number and speed of carriers in membranes

1. **What type of molecules are carriers?** Our ignorance about the nature of carriers is only too well exposed by the fact that we have been able to write so much about them without once answering the question posed as the title of this section! We know very little about the nature of the carrier molecules. With the few facts at our disposal and some intelligent guesses we can suggest a possible structure. Firstly, they are likely to be mainly protein in nature and thus of fairly large molecular weight. They must have special sites that can bind solute molecules and these sites will display a high degree of specificity. The actual area of the binding site will be a minute proportion of the total area of the carrier molecule. In all these features carrier molecules resemble enzymes. Both are proteins, both have special binding sites and both are blocked or inhibited by heavy metal ions and chemical reagents that attack groups present in proteins. While nearly a thousand different enzymes are known it is unlikely that as many carriers will be uncovered. Various solutes probably use one carrier but their affinities for this carrier will vary enormously.

Research on isolating carriers from membranes is progressing along a number of different paths. One method utilizes the fact that carriers ferry mainly lipophobic solutes across lipid membranes. Thus if a carrier is isolated it should be able to bind its normal lipophobic solute and make it "soluble" in lipoid phases. A number of research laboratories are hard at work trying to isolate specific carrier molecules from bacteria and red cell membranes. Usually they extract a fraction of the membrane that binds a solute known to have a carrier mediated transfer into the cell. Some work suggests that the isolation of pure carrier molecules from bacteria is not very far away. Serious difficulties exist, however, just as they do in the extraction of structural proteins from membranes. The basic problem being whether the extracted substance that binds the solute is the same as the carrier molecule in the cell membrane. Some scientists preach a sermon of doubt, arguing that the whole point of membrane carriers is that they are intimately bound up with the structure of the membrane: break this down and you destroy the carrier. This may be partly true but it can also be said in defence that the science of biochemistry rests on people brave enought to grind, smash, tear and disrupt cells into their component parts. Membrane workers should be as dauntless as their cytoplasmic colleagues!

2. **How many carriers are there in the cell membrane?** We have shied away so far from giving a figure for the number of carrier

TABLE 11. *Estimates of the number of carriers for glucose in cell membranes (after Stirling 1967)*

| Cell type | Number of carriers per cell | per square micron membrane |
|---|---|---|
| Mucosal cell, small intestine (hamster) | $26 \times 10^5$ | 1,700 |
| Tubule cells of the kidney (dog) | $100 \times 10^5$ | 6,700 |
| Red blood cell (human) | $14 \times 10^5$ | 11,400 |
| Red cell ghosts (human) | $5 \times 10^5$ | 4,100 |

molecules we would expect to find in a cell membrane for a specific transfer process. This is a difficult measurement to make but there are now a number of studies on the carriage of glucose across cell membranes for a reasonable estimate to be made. Various methods have been adopted. One approach is to see how much glucose is bound by a known number of cells while another measures the uptakes of inhibitors that are thought to be bound only to the glucose transfer sites. The total number of glucose carrier sites in red cells, kidney tubule cells and mucosal cells of the small intestine have been estimated by these techniques. The data obtained are shown in Table 11. The number of carriers present vary from the lowest figure of $5 \times 10^5$ for red cells up to the highest of $1 \times 10^7$ for a cell from a dog kidney tubule. If the estimate for the number of carriers for each cell type is divided by their respective areas (a figure which is itself an approximation) the number of carriers per square micron of cell surface can be obtained. These range from the high density on red blood cells (11,400) to the low density on hamster small intestine (1,700). Obviously because these calculations of carriers per unit area are based on two estimates which are then divided, large errors may be present. Even so, they give us some idea of the magnitude involved. What they do not tell us is how much of the cell surface is taken up by the carriers. If we arbitrarily assume that each carrier is a sphere of 100 Å diameter (the thickness of the plasma membrane) then the maximum fraction of the cell surface occupied by the carrier would range from 13 per cent for the small intestine up to 89 per cent for the red cell. This latter calculation cannot be thought of as a "real" value but only as an index of what is possible.

3. **How fast do carrier molecules handle solute transfer?** With a carrier molecule carrying glucose back and forth across the membrane a value that can be calculated is the number of molecules of glucose the carrier can handle per second. In enzymes this is called the turnover number, there is no reason why we cannot also use this term for carriers. The red cell carrier has been estimated to have a turnover number of about 180–500 glucose molecules/carrier/sec. Those of the intestine and kidney are much lower, at around 21 and 25 respectively. These turnover numbers for carriers compare favourably with those obtained for enzymes.

Another way of thinking about the speed of carriers is to calculate how fast they move through the membrane compared to a molecule moving through the cell cytoplasm. Although this sort of calculation uses a lot of assumptions it leads to a very interesting result. We know that it takes approximately 5 seconds for a glucose molecule to cross the kidney tubule cell when it is absorbed from the tubular fluid into the blood. Of this time, it is calculated that the glucose spends a maximum of 0·04 seconds crossing the membrane with the carrier. Now although this may seem a short time, being only $\frac{1}{25}$ of a second, it should be remembered that the distance traversed is only some 100 Å. Using this distance and the time taken we can calculate by a simple formula the Diffusion coefficient of the glucose–carrier complex. It comes to $4·2 \times 10^{-12}$ cm$^2$/sec. In comparison the diffusion coefficient for the movement of glucose through the rest of the cell is $0.2 \times 10^{-6}$ cm²/sec. It appears then, that the passage of glucose through the cell cytoplasm is much faster than that through the membrane via the carrier. In fact the movement of glucose via the carrier is some 50,000 times slower than its movement through the entire cytoplasm! (Diedrich 1966). This dramatically illustrates the concept of the membrane as a barrier to diffusion.

# CHAPTER 7

# The Movement of Charged Solutes

Many substances that are of great importance to the daily life of cells are charged. Although examples of charged solutes abound, the most important in many ways are the ions potassium, sodium and chloride. While many of the mechanisms that transfer neutral molecules across cell membranes can also handle charged molecules, the movement of charged substances usually entails the generation of an electrical potential difference across the membrane. Such a potential difference will itself affect the movement and distribution of other charged solutes; negative ions would be attracted to the positive side of the membrane while positive ions would tend to be repelled. The interactions between electrical potentials and the movements of charged substances make the membrane transfer of these substances much more complicated than those of neutral molecules.

## The traffic in ions

Practically all cells, whether they are fixed in tissues or are free to wander, have been found to have an intracellular ionic make-up very different from the solutions in which they live or are bathed (Table 12). Most mammalian cells have a high intracellular potassium ion concentration (approximately 140 millimolar) and a fairly low sodium concentration (approximately 20–40 millimolar). The tissue fluid that bathes mammalian cells contains approximately 145 millimoles $Na^+$ per litre and about 5 millimoles $K^+$ per litre. When such cells are killed, or their metabolism is reduced to low levels, they gradually accumulate sodium and lose their potassium, two features which might well be expected from the normal concentration gradients of these ions. Before the advent of radioactive tracers of sodium, potassium and chloride the general opinion was that ions did not readily penetrate cell membranes and that cells were "born" with their peculiar ionic make-up and remained thus for the rest of their lives. The reason for this gross error was the difficulty in measuring, with any accuracy, the rapid influx and efflux of ions. Experiments with the newly made radioactive ions of sodium and potassium in the 1940s and the development of flame photometry for accurately measuring sodium and potassium ions in solutions allowed both individual and net ion movements to be identified and detailed. Within a few years it was clearly established

TABLE 12. *Concentrations of potassium, sodium and chloride in the blood cells and environmental bathing fluids of various species (after Prosser* et al. *1952)*

| Animal | Fluid or cells | $K^+$ | mM/Kg $Na^+$ | $Cl^-$ |
|--------|----------------|-------|--------------|--------|
| Man | blood plasma | 5 | 143 | 103 |
| | red blood cells | 105 | 10 | 80 |
| Frog | blood plasma | 3 | 104 | 74 |
| | muscle | 85 | 24 | 11 |
| | sea water | 10 | 476 | 562 |
| Squid | blood plasma | 17 | 354 | 469 |
| (Loligo) | Giant axon (axoplasm) | 310 | 44 | 130 |
| | muscle | 114 | 54 | 71 |

that there was a huge traffic in ions in every cell that was investigated. Today we know that cells in general pump potassium into and sodium out of their cytoplasm. This pumping is continuous, the energy for the work done is thought to come from the hydrolysis of high energy phosphate compounds like ATP. These are formed from the enzymatic breakdown of foods via the metabolic pathways.

## The sorption theory of ion accumulation

It is perhaps relevant at this point to mention a theory of cellular ion accumulation that does not use the concept of ion pumps in the cell membrane. Some modern biologists have argued that the different ion concentrations between cells and their environment are due not to the perpetual pumping at the cell membrane of potassium in and sodium out, but to the ability of the intracellular matrix of proteins and other large non-diffusible structures to preferentially bind potassium rather than sodium. The inside of the cell is thus like a giant sponge the "holes" being filled with bound or "sorped" potassium ions. The theory is generally given the name "sorption theory". Two important experimental facts are unexplained by such a theory. Firstly, it is known that in a number of cells the mobility of potassium ions inside is nearly the same as that measured outside in free solution (see also page 100). Clearly if a large proportion of the potassium ions were bound up with negative charges on protein or other large molecules there would be a restraint on their intracellular mobility. The second experimental fact is that although potassium can move about inside the protoplasm of some cells with apparent ease, if it is placed on the outside it enters with difficulty. There clearly appears to be a barrier between the inside and the outside of the cell, i.e. a functional membrane.

### Active and passive movements of ions

Just as the movement of neutral molecules can be divided into active and passive transfer, so can that of ions. Active ion transfer occurs when the ion in question is moving against both its chemical and electrical gradients. In practical terms this means that negative ions have to be moved against their chemical concentration gradient and a negative electrical potential difference, while positive ions have to be transferred against a positive electrical potential difference and of course their chemical concentration gradient.

Passive transfer of ions occurs when ions "free-wheel" down either a chemical concentration gradient or a favourable electrical potential difference. When a membrane has a potential difference across its structure, negative ions will tend to move to the positive side and positive ions to the negative side. The rate of accumulation will depend on how permeable the membrane is to particular ions (often called its "conductance" to the particular ion), and the numerical value of the potential difference. As there is a relationship between p.d. and ion movement we can in fact fairly accurately assess how much ion will accumulate across charged membranes. Because such quantitative measures are vitally important in our understanding of the biological mechanisms for ion movement we must take a quick look at some of the relationships between the movements of ions and the potentials generated and between potentials and the consequent movement of ions they cause.

### The physical chemistry of ion movements across membranes

1. **Ion movement across a simple membrane – the flux concept.** Let us take a membrane that allows small solute molecules and water to permeate but prevents the passage of large ions of colloidal size (Fig. 39). If we add potassium chloride solution to both sides of the membrane no electrical potential difference will be generated across the membrane because both the potassium and the chloride

ions will diffuse through the membrane in equal amounts in both directions, A to B and B to A. This means that for each ion there is a movement from A to B and from B to A. Biologists call these movements "one way or unidirectional fluxes", the flux of a solute being defined as the total number of permeant molecules crossing a unit area of membrane in unit time. There is thus a potassium flux from A to B and one from B to A. At equilibrium these two fluxes must be identical. Hence there is no net movement (net flux) of potassium in the system but there are of course always an enormous number of potassium ions diffusing between A and B. A similar situation exists for chloride ions, the unidirectional fluxes of chloride from A to B and B to A are large but there is no net chloride flux. Because it is often inconvenient to keep using terms like "the one way flux of potassium from compartment A into compartment B", a simple shorthand is used. A flux is abbreviated as J and the compartment that the movement takes place from is placed as the first subscript and that it enters is placed as a second subscript. Hence our sentence "the one way flux of potassium from A to B" becomes $J_{AB}$. Under normal circumstances $J_{AB} = J_{BA}$ both for potassium and chloride. The two solutions have no electrical potential difference between them as at any particular time each solution must be electrically neutral, i.e. contain as many potassium as chloride ions.

What happens if we make the membrane permeable only to one ion species, in technical jargon, what happens when the membrane becomes permselective to potassium ions only?

### Electrical potentials across a permselective membrane

If the membrane only allows potassium ions to pass then some will diffuse through without chloride ions, but as the electronegative chloride ion has a very strong attraction for the electropositive potassium ion, there will be an enormous restraint on these diffusing potassium ions. The electrostatic attraction of chloride ions will prevent much potassium ion movement. If we had equal concentrations of potassium chloride on both sides of the membrane the total potassium flux $J_{AB}$ would clearly be the same as $J_{BA}$ hence there would be no potential difference generated across the membrane (Fig. 39, 2). But if we place in compartment A a solution of potassium chloride that is ten times more concentrated than one in compartment B we have now given the potassium and chloride ions a chemical concentration gradient down which they can diffuse.

Clearly if there were freedom of diffusional movement the net flux
of both potassium and chloride would be from A to B. Because the
membrane will allow only potassium ions to move through its

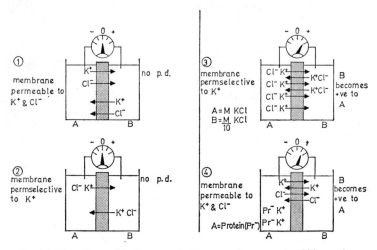

*Fig.* 39. The effects of ionic concentration, membrane permeability and non-
diffusible negative ions on the development of an electrical potential difference
across an artificial membrane.

1. In this experiment the membrane is permeable to both $K^+$ and $Cl^-$ ions
and there are equal concentrations of KCl in compartments A and B. No
potential is generated.

2. The membrane is now made permselective to potassium ions only but
equal concentrations of KCl are still maintained in A and B. No potential is
generated.

3. The membrane is made permselective to $K^+$ ions only. Molar KCl is
placed in compartment A while 1/10M KCl is placed in compartment B. A
potential difference is generated, B becomes positive to A.

4. The membrane is made permeable to both $K^+$ and $Cl^-$. Compartments
A and B initially have equal quantities of KCl but a small amount of protein
(Pr⁻) is placed in A as well. This results in an equilibrium being set up (Gibbs–
Donnan) between the ion concentrations and the potential. B becomes positive
to A.

structure then only these ions can flow from A to B. As soon as one
potassium ion leaks through the membrane without its accompany-
ing negative chloride ion, the membrane becomes electrically
polarized, an electrical potential difference is generated across the
membrane such that compartment B becomes positive to A

(Fig. 39, 3). The next potassium ion diffusing across will have even greater difficulty in moving from A to B because it is leaving a negatively charged chamber for a positively charged chamber, and so on. The important question that arises from this type of system is, what is the quantitative link between the magnitude of the potential generated and the concentration gradient of the ions in the two compartments?

### The Nernst equation – the link between concentration and electrical potential

The relation between the concentration of ions and the value of the potential generated by the movement of the ions down their chemical concentration gradient was formulated by Walter Nernst in 1889 when he was twenty five years old. The equation had to bring together electrical potentials and concentration gradients in the same units. Because of this the equation took a somewhat complex form. Nernst found that when a univalent positive ion was moving through the membrane the potential difference,

$$E = \frac{RT}{F} \log_e \frac{\text{conc. ion side A}}{\text{conc. ion side B}}$$ while if a univalent negative ion was penetrating the membrane then, the potential difference

$$E = \frac{RT}{F} \log_e \frac{\text{conc. compartment B}}{\text{conc. compartment A}}.$$ In all these equations R = the gas constant, T = temperature in absolute units and F = the Faraday (96,500 coulombs, or the charge in coulombs on one ion mole of a univalent ion). If we convert the natural logs (logs to base "e") to base 10 and calculate $\frac{RT}{F}$ for a univalent ion at 20° C, the equation simplifies to $E = 58 \log_{10} \frac{\text{conc. side A}}{\text{conc. side B}}$ for a univalent position ion. For a univalent negative ion, as the potential generated must be of reversed sign, the concentrations in the equation are reversed so it becomes $E = 58 \log \frac{\text{conc. side B}}{\text{conc. side A}}.$ In both equations E is in millivolts and the potential measured is B with respect to A.

### The use of the Nernst equation in characterizing biological membranes

The Nernst equation has been of enormous value in characterizing the ionic permeability of biological membranes. The simple way it can be applied is shown by an example using the frog skin "membrane". It has been known for a hundred years that the inside of frog skin is positive compared to the outside. Though the finer points of the ion movements that cause the potential to be developed

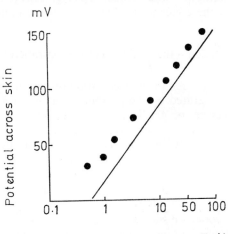

*Fig.* 40. Effect of changing the sodium ion concentration of the fluid bathing the outside of a frog-skin membrane on the electrical potential difference across the skin. The dots represent observed values while the solid line represents the theoretical change that would occur if the outer membrane of the skin was permeable only to sodium. It was calculated from the Nernst equation and has the theoretical slope of 58 mV for a tenfold change in sodium concentration. The close fit between the slope of the theoretical line and that of the frog skin shows that the outer membrane behaves as a permselective sodium membrane (i.e. sodium electrode) (from Koefoed-Johnsen and Ussing 1958).

are still under discussion it appears that the transfer of sodium from the outside of the skin to the inside is responsible for part of the electrical activity. Hans Ussing and Valborg Koefoed-Johnsen found that the electrical potential measured across the tissue varied with the concentration of the sodium ion. When they plotted the $\log_{10} \dfrac{\text{Na conc. outside the skin}}{\text{Na conc. inside the skin}}$ against the skin potential difference, the resultant graph was a straight line (Fig. 40). This indicated that the outer membrane of the skin behaved as if it was permeable mainly to sodium ions. We say that the outer membrane of the frog skin is behaving like a sodium electrode because the potential developed is dependent on the concentrations of the sodium ions across its structure. If the range of sodium concentrations chosen is tenfold (say from 5 to 50 millimolar, or from 10 to 100 millimolar sodium) the slope of the line at 20° C will always be 58 mV. This is because the Nernst equation contains the term $\log_{10} \dfrac{\text{outside conc. Na}}{\text{inside conc. Na}}$. When we arrange the two concentrations so that one is ten times the other we are taking the log to base 10 of the number 10. Now a log is defined as the power to which the base has to be raised to get the number, thus $\log_{10} 10 = 1$. Hence in the Nernst equation the potential difference produced by a tenfold difference in concentration of a permeant ion will be some 58 mV at 20° C. By this simple test we can tell if a biological membrane is acting like a permselective membrane for any particular ion. If we change the concentration of the permeant ion and get a slope near to the theoretical 58 mV for a tenfold change in concentration, the membrane can be said to act as if it were permeable mainly to that ion species, i.e. as an electrode of that ion.

While we have used the outer membrane of the frog skin as an example of a biological permselective membrane for sodium ions we could have chosen the membrane of nerve or muscle cells. In their resting states both these membranes behave as if they were permselective to potassium ions, i.e. as potassium electrodes. Changing the external concentration of potassium on the outside of these cells alters the potential across the cell membranes (measured by an intracellular electrode arrangement as in Fig. 9). Increasing the external potassium decreases the membrane potential. A tenfold change in the external potassium concentration alters the potential by some 58 mV at 20° C. Other cell membranes show only a partial relation between potential and external potassium concentration presumably because other ions permeate the membrane.

### Electrical potential differences caused by diffusion of ions – diffusion potential differences

All ions have specific speeds of movement through water, some moving faster than others. Thus if two ions in a liquid are moving into a region that is not as concentrated as that from which they are leaving, a potential difference will occur if one ion has a greater mobility than the other. In biological fluids although many ions are present, sodium, potassium and chloride tend to dominate the scene with regard to the development of potentials. The mobilities of potassium and chloride are very nearly the same but that of sodium is lower than chloride or potassium. Hence a concentrated solution of NaCl in contact with a more dilute one will create conditions to cause a potential difference to be generated between the two fluids. This is because the chloride ion diffuses faster into the dilute solution than the more sluggish sodium ion, the concentrated solution thus becomes positive in relation to the dilute solution. By a modification of the Nernst equation to include mobilities of the ions we can characterize the diffusion potential generated by a salt moving from concentration $C_1$ to concentration $C_2$ by the equation,

$$E = \frac{RT}{F} \frac{(u - v)}{(u + v)} \log_e \frac{C_1}{C_2}$$

where u is the mobility of the positive ion, v is the mobility of the negative ion and the other symbols are as in the Nernst equation on page 113.

Diffusion potentials will occur not only in solutions containing ions of different mobilities but also across membranes where one ion species migrates faster than another. Clearly, if a membrane allows a negative ion through faster than a positive ion, a diffusion p.d. will occur. As biological membranes contain charged molecules (phospholipids and proteins have electrically charged groups) such membranes would be expected to hinder one type of ion species more than another.

If the membrane's pores had a large number of fixed negative charges then they would hinder negative ions but allow positive ones through fairly readily (Fig. 41). On the other hand, if the membrane was positively charged it would allow negative ions through rather than positive ones. Thus if we experimentally set up a concentration gradient of a salt containing an anion and a cation of physiological importance (say sodium chloride) across a biological membrane and measure both the electrical potential and the polarity developed by the salt diffusing across the membrane we should be able to tell whether the membrane has a net negative charge or a net positive one in its pores. If it were negatively charged the membrane would let sodium ions diffuse through somewhat faster than chloride ions (as both are moving down their concentration gradients from a high to a low concentration). The solution that was the most dilute in sodium chloride would become positively charged. If, however, the membrane was positively charged, the chloride ions would diffuse through faster than the sodium ions

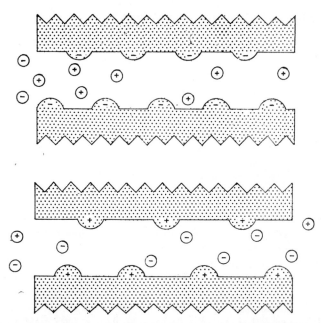

*Fig.* 41. Representation of two charged pores in a cell membrane. The top one, a pore with a preponderance of negative charges in the membrane matrix, will tend to be filled with positive ions the negative ions being repelled and excluded. The lower one has a preponderance of positive charges in its matrix, hence it is filled with negative ions and excludes positive ones. Both types of pore may exist in cell membranes but if there are more negative than positive charged ones, the membrane will behave primarily as a negatively charged one.

and the dilute solution would become negatively charged. This sort of experimental approach has been tried out on a few biological membranes. The mucosal cells of the small intestine are thought to have negatively charged pores because they let sodium ions through slightly faster than chloride ions. Of course one can correlate the actual value of the diffusion potential, generated by various concentrations of sodium and chloride, with the Nernst equation. When

E

this is done with the intestine it appears that the luminal membranes act only as a partial sodium electrode, i.e. they are only partially permselective to sodium ions. In this situation a tenfold change in the sodium chloride concentration does not give the full 58 mV potential difference change across the membrane. This is interpreted to indicate that the luminal membrane of the cells are permeable to ions other than sodium.

**A further modification of the Nernst equation for biological membranes.** We should briefly mention that the modification of the Nernst equation on page 116 that allowed calculation of diffusion potentials in solution for a single anion and cation cannot unfortunately be applied to potentials generated across biological membranes by a number of ions. In this case the various ions have different mobilities (or permeabilities as they are usually called) in the membrane and a more complicated version of the Nernst equation is necessary. Membrane permeabilities of ions are usually designated by the symbol P with a subscript to denote the ion. Thus $P_K$ is the permeability of potassium ions in the membrane, $P_{Na}$ that of sodium ions and $P_{Cl}$ that of chloride ions. It has been found possible to link up the potential with the permeability and the internal and external concentrations of anions and cations across the nerve axon membrane when certain conditions are assumed. Even under these specific conditions the equation is a formidable one and is only given below to indicate the complexities of membrane potentials and ionic movements. The equation was first developed by D. Goldman in 1943 and his name is usually attached to it as a label. The potential (E millivolts) across the nerve axon membrane was found to be given by

$$E = \frac{RT}{F} \log_e \frac{P_K [K_0] + P_{Na} [Na_0] + P_{Cl} [Cl_1]}{P_K [K_1] + P_{Na} [Na_1] + P_{Cl} [Cl_0]}$$

The subscripts 1 and 0 on the ions are to indicate the inside and outside concentrations respectively. When the experimental values for these various permeabilities and concentrations are slotted into the Goldman equation the calculated potential across nerve axons agrees with that found experimentally.

### Potential differences across cell membranes resulting from the Gibbs–Donnan equilibrium

Many macromolecules present inside cells, like proteins and polyphosphates, carry net negative charges yet cannot pass through cell membranes. If the membrane is permeable to ions it is obvious that the negative charges on the non-diffusible molecules will affect the distribution of ions across the membrane and create an electrical potential. The theory of this type of system was worked out by Willard Gibbs at Yale in Connecticut in the late 1870s but was investigated in the laboratory by Frederick Donnan in 1924. The equilibrium between the ions and the potential difference generated by this system is thus known as the "Gibbs–Donnan" equilibrium. We can use again the simple model of the membrane separating the two compartments A and B to see what happens when protein is added to the system. Initially, if we have equal concentrations of KCl on both sides of the membrane (as in Fig. 39) there is no electrical potential across the membrane. If we add a small amount of negatively charged protein to one compartment, say to A, then a number of potassium ions will be attracted to balance the negative charges on the protein molecules (Fig. 39, 4). This must cause a rearrangement of diffusible chloride ions across the membrane. The total amount of potassium ions in A (the side containing the protein) will increase, and become greater than the chloride ion concentration, while the amount of potassium and chloride in compartment B will be altered but their concentrations will equal one another. This rearrangement of ions sets up a potential difference across the membrane because of the imbalance of potassium ions between A and B. They are always trying to leak from their high concentration in A down their concentration gradient to B. Thus B becomes positively charged with respect to A. It should be obvious that although the system is in equilibrium the concentrations of ions in A and B will never be equal. We can in fact calculate what these concentrations will be. As the system only has one potential difference across the membrane (E millivolts) all the diffusible ions must be in equilibrium with this potential. Now the Nernst equation makes it possible for us to relate the potential to the ion concentrations

Thus for the potassium ion, the potential difference between A and B,

$$E_1 = \frac{RT}{F} \log_e \frac{\text{conc. [K] in A}}{\text{conc. [K] in B}}$$

similarly for chloride, the potential difference between A and B,

$$E_2 = \frac{RT}{F} \log_e \frac{\text{conc. [Cl] in B}}{\text{conc. [Cl] in A}}$$

As there is only one potential across the membrane between A and B, $E_1$ must equal $E_2$, thus

$$E = E_1 = E_2 = \frac{RT}{F} \log_e \frac{[K]_A}{[K]_B} = \frac{RT}{F} \log_e \frac{[Cl]_B}{[Cl]_A}$$

or

$$\frac{[K]_A}{[K]_B} = \frac{[Cl]_B}{[Cl]_A} \therefore [K]_A [Cl]_A = [K]_B [Cl]_B$$

The latter equation states in mathematical terms that the product of the diffusible ions in each of the compartments must be equal to each other. It can be used to calculate the amounts of ions in each compartment at equilibrium.

Practically all cells have a Gibbs–Donnan equilibrium across their membranes because these are permeable to ions but not the large macromolecular intracellular anions. Although the potential set up by this type of equilibrium of diffusible ions is often called a Donnan potential it is of course still quantified by the Nernst equation. While it is true that all cells have non-diffusible anions their intracellular concentration is too small to be responsible for the high intracellular potassium concentration and the usually high membrane potentials. The major cause of both these is the presence of ion pumps in cell membranes. We must now turn our attention to the actual mechanisms that pump ions into and out of cells.

### The ion pump in cell membranes

As we stated previously, it is now widely accepted that the high concentration of potassium and low concentration of sodium inside most cells depends on the activity of a pump which is thought to be located in the cell membrane. The machinery of the pump must expel sodium ions that are constantly leaking into the cell and pump back potassium ions that are constantly leaking out (Fig. 42). Experiments have shown that removal of potassium from solutions bathing cells interferes with the extrusion of sodium while suppression of sodium extrusion interferes with the uptake of potassium. Such findings have given rise to the idea that the active movement of sodium and potassium is by a coupled pump, so many sodium ions extruded from a cell for so many potassium ions taken up. In technical terms there appears to be a "stoichiometry" between the uptake of potassium and the extrusion of sodium. The stoichiometric relation between the uptake of potassium and sodium extrusion varies in different cells. Using red cell ghosts and membranes from giant axons it has been calculated that every time the "pump" turns over three sodium ions are expelled for every two potassium ions that are pumped in. The ion pump in frog-skin epithelial cells, however, is thought to have a one for one stoichiometry. Every time one potassium is pumped in one sodium is

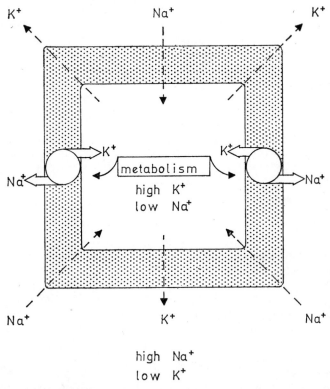

*Fig.* 42. A hypothetical general cell showing the pump-leak model of ion regulation. Cell metabolism supplies the energy for the continuous working of the $Na^+$–$K^+$ linked ion pumps thought to be located in the cell membranes . As the intracellular $K^+$ is higher than the extracellular concentration, $K^+$ ions are constantly leaking out of the cell (dotted lines). With $Na^+$ ions the picture is reversed, they are constantly leaking into the cell (dotted lines) because their concentration is greater outside the cell than inside. When metabolism is reduced or prevented by metabolic poisons, the ion gradients maintained by the pumps run down.

expelled. If we assume that the supply of energy for the movement of the sodium and potassium ions comes from the hydrolysis of ATP, we would expect that there should be an enzyme in, or bound close to, the membrane site where the ions are transferred. This enzyme should be able to hydrolyse ATP. Now although many enzymes were known that could break down ATP by hydrolysis, none were known to be present in the membrane fraction of cells until the work of Skou in 1957. He found that particles from minced crab nerves could hydrolyse ATP. More important, however, was the observation that while this enzyme activity was enhanced by sodium ions, the addition of potassium ions in the presence of sodium ions enhanced the ATP splitting activity even more. Because of the observation that the ATP splitting action was enhanced by sodium and potassium ions, Skou suggested that the enzyme's activity was in some way related to the ion transport of cell membranes. The enzyme was called adenosine triphosphatase or "ATPase" for short. Since the original studies in 1957, an enormous amount of research has been directed at trying to understand the possible link between this sodium–potassium activated ATPase and the ion pumps in cell membranes.

### The $Na^+$ – $K^+$ activated ATPase as an ion pump

Practically all cells that have been homogenized have a $Na^+$ – $K^+$ activated ATPase in the membrane fraction of the homogenate. The activity of the enzyme is usually related to the amount of cation pumping that has to be done; high activity is found in cells with a high potassium concentration and low activity in cells with a low intracellular concentration. Specific inhibitor substances well known to prevent cells pumping potassium in and sodium ions out (a range of substances known as cardiac glycosides of which ouabain is the most often used) also inhibit a large part of the $Na^+$ – $K^+$ stimulated ATPase activity. Because of this, it is thought that the ouabain-sensitive ATPase is the fraction of the enzyme activity responsible for normal ion pumping. The most dramatic evidence, however, for a connection between the ouabain-sensitive $Na^+$ – $K^+$ ATPase and cation transfer has been observed using red cell ghosts.

1. **ATPase as the ion pump in ghosts.** Because red cells can be reversibly haemolysed (page 29) it is possible to remove all their normal intracellular constituents and replace them with simple solutions containing basic ions and some ATP. If the cells are then

made to reseal themselves they are capable of pumping potassium from the outside bathing solution across the membrane into their new artificial intracellular solution. This type of experiment has shown that the ATPase in red cell membrane is so orientated that it needs $K^+$ ions on the outside surface and sodium ions on the inside surface to be efficient in pumping cations across the ghost membrane. It seems that the ATPase in the red cell membrane has an asymmetry or a "vectorial" factor compatible with its function as a pump for the alkali metal ions.

2. **ATPase as the ion pump in squid giant axons.** Because squid giant axons are large, it is possible to microinject various substances into their axoplasm without damaging the axon membrane. This technique was used to show that ATP was needed to power the sodium pump in the axon membrane. First the axon was poisoned with cyanide, a metabolic inhibitor that interfered with normal production of ATP from metabolism. Then, when the sodium extrusion from the axon had fallen to low levels (as the ATP inside the axoplasm had run down) fresh ATP was injected into the axoplasm by means of a micropipette. The extrusion of sodium rapidly approached that in the normal axon. External application of ATP to the axon membrane was found to have no effect. As there is an ATPase in the squid axon membrane it is highly likely that the ATP was being hydrolysed by this enzyme and thus supplied the energy for the sodium ion extrusion.

Although many other experimental facts show a parallelism between ATPase and ion transfer, we are far from an intimate, detailed knowledge of how the enzyme actually transfers the potassium and the sodium ions across the membrane. In fact we really do not know whether ATPase is a single enzyme or an enzyme system or even whether ATPase is involved in the actual transferring function. The enzyme(s) could simply be involved in a linked sequence of reactions that change the properties of part of the membrane which then transfers the sodium and potassium rather than being the actual carrier. Whatever the ultimate mechanism, any model that incorporates the transfer of potassium for a sodium ion has to involve binding sites for these ions that are alternately exposed to the inside and the outside of the cell membrane. What possible molecular arrangement for such a system can be envisaged? In a simple question, what does the $Na^+ - K^+$ pump look like?

## A possible model for the membrane ATPase

At the present moment we can only speculate on the molecular architecture of the ATPase ion pump. With further experiments our ideas about ATPases and cation transfer will undoubtedly undergo a radical revision. Even so models have been devised that suggest the ATPase as a large protein unit built into the membrane, forming a structure that is made to transfer sodium and potassium through

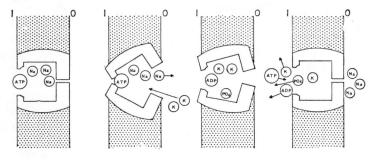

*Fig.* 43. The $Na^+$–$K^+$ ATPase ion pump according to the model of A. Lowe (1968). The ATPase pump stretches across the membrane from the inside (I) to its outside (O). When it is closed (first diagram on left) only the intracellular ions have access to the binding sites. When the sodium ions and the ATP fit on to their sites the hemispherical structure closes exposing the sodium ions to the outside fluid. The closing of the structure makes the sites less favourable for sodium and potassium ions displace them and bind to the sites. As soon as the potassium ions are attached the ATP is hydrolysed to ADP, by removal of a phosphate group ($PO_4$). When this happens the structure cannot remain open and it closes exposing the potassium ions to the inside of the cell. The affinity of the site binding the ions again changes, the potassium diffuses away from the pump structure into the cell, together with the ADP and phosphate. Fresh ATP and sodium ions move into the pump, become attached and the structure again closes. In one complete pump cycle, three sodium ions are pumped out, two potassium ions pumped in and one ATP molecule is hydrolysed to ADP and phosphate.

the membrane by the agency of ATP. The most recent model uses a pair of rotating hemispheres, which are part of the enzyme that hydrolyses ATP (Fig. 43).

When the molecular arrangement is open on the inside of the cell it exposes three binding sites for sodium and one for ATP. When these are filled the hemispheres close and automatically open the inside of the structure to the outside of the cell and the ion binding sites become less attractive for sodium but more attractive for potassium. The sodium ions diffuse away and two potassium ions become attached to the sites. This causes the ATP to be hydrolysed to ADP and phosphate ions and the hemispheres close.

Once exposed to the intracellular environment of the cell, the intracellular sodium ions compete for the potassium sites while fresh ATP effectively displaces the ADP which diffuses into the cell interior with the phosphate and is remade to ATP. As soon as the sodium ions and ATP become attached, they activate the closing of the hemispheres and expose the sodium ions once again to the outside solution. The whole process is continuously repeated, pumping three $Na^+$ ions out and two $K^+$ ions in for every ATP molecule hydrolysed. The specific inhibitors of ion pumping processes, the cardiac glycosides like ouabain, probably act by binding on to the ion specific sites when they are exposed to the outside of the cell. The hemispheres are thus kept open and are effectively locked, exposed to the outside medium only. Cells poisoned by cardiac glycosides will slowly lose their potassium and gain sodium owing to the run down of the concentration gradients usually maintained by ion pumping.

Although this model ATPase-ion pump is in no way proved as being identical with pumps thought to exist in cell membranes, it does illustrate the essential features of sodium and potassium ion transfer across such membranes. These features are:

1. Specificity for $Na^+$ and $K^+$ exchange.

2. Stoichiometry between $K^+$ and $Na^+$ ions.

3. Pumping the ions uphill against a concentration gradient (active transfer).

4. Enzymic nature of pumps in cell membranes.

5. Asymmetric nature of ion requirements for enzymic ATPase activity in membranes.

6. Inhibition of ion pumps by specific inhibitors – the cardiac glycosides like ouabain.

7. ATP usage for osmotic work (i.e. ion transfer).

### Electrogenic ion pumps

If a carrier accepts an ion, transports it across a membrane and releases it on the other side as a free ion, then clearly a transfer of electrical charge has taken place and an electrical potential difference will be directly built up across the membrane by this pumping device. The question with biological pumps is whether they too can transfer charges across membranes to directly build up electrical potential differences in this manner. This is a very vexed question and in many respects has not yet been satisfactorily answered. The difficulty is that, as we have seen, there is usually more than one mechanism in operation generating electrical activity across cell membranes. Potentials across membranes arise from:

1. Gibbs–Donnan equilibrium potentials (Nernst relationship).
2. Diffusion potentials.
3. Active ion transfer.

Any experiment made to elucidate one potential generating mechanism automatically interferes with the others. This interaction makes it far from simple to disentangle the processes of ion transfer and electrical activity.

There are many investigators who think that there is evidence that biological pumps can produce electrical potentials by directly transferring ions from one side to another – they call these pumps "electrogenic pumps". Evidence for such pumps has been found in a number of cells but probably the most dramatic indication of their operation is seen in the sheets of epithelial cells that line the alimentary and reproductive tracts of man and animals. These epithelia are in reality sheets of cells working in concert, transferring ions from the lumen into the blood and/or from the blood into the lumen. They, of course, have to transfer the ions across two cell membranes; the one in contact with the blood or tissue fluid and the other in contact with the fluid in the lumen of the organ.

### The electrogenic sodium pump of the small intestine

A fascinating electrogenic ion pump found in an epithelial cell is the sodium pump in the base of the mucosal cell of the small intestine. Such cells have two membranes. One is exposed to the lumen and has the peculiar "coupled carrier" for sodium and glucose entry, the other membrane is at the blood side of the cell and possesses the actual sodium pump (Fig. 38). When the lumen of the intestine contains sodium ions but no free glucose, the potential difference measured across the cells is small, less than 3 millivolts, lumen negative to blood (Fig. 44). If some glucose is added to this sodium containing solution bathing the luminal membranes, the potential difference rises rapidly so that within seconds it reaches 10 millivolts or more. The reason for this rapid increase in electrical potential is thought to be related to the increased entry of sodium that occurs when glucose and sodium enter on the coupled carrier. This carrier moves only when both solutes

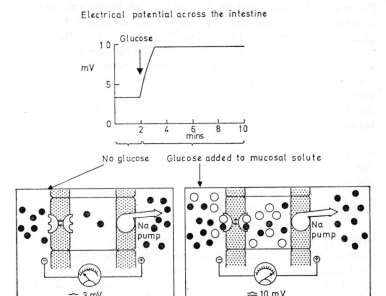

*Fig.* 44. The electrical potential across the mucosal cells of the small intestine and the effects of addition of the actively transferred sugar, glucose. This composite diagram shows what happens when glucose is added to the fluid bathing the mucosal surface of a piece of intestine (upper graph). In the absence of glucose there is a low potential across the gut wall of about 3 mV, the lumen being negative to the blood. This potential is probably caused by small amounts of sodium ions (solid circles lower left diagram) which leak into the mucosal cell being pumped out by the sodium pump. Sodium ions enter either by diffusion or a few can move across by the carrier (shown empty). Addition of glucose (large open circles) to the mucosal fluid causes a very rapid increase in the potential (upper graph) which reaches a new steady state of about 10 mV in a few minutes. This increased potential is thought to be due to the increased entry of sodium into the mucosal cell via the "glucose–$Na^+$-carrier" complex facilitating rapid movement of the carrier across the mucosal membrane. Hence more sodium becomes available to the $Na^+$ pump. It thus pumps more sodium out of the cell, creating a greater electrical potential. At the same time the cell accumulates glucose molecules which then diffuse out into the blood stream. Active transport of sodium has thus given rise to accumulation or active transfer of glucose.

are attached to the sites. Thus without glucose the sodium pump is "starved" of sodium and the potential is low because the amount of positive charge transferred to the serosal side is low. As soon as glucose and sodium enter the cell via the coupled carrier the sodium pump has access to more sodium and thus up goes the potential difference. In other words the mucosal membrane is a rate limiting membrane for the entry of sodium while glucose apparently increases its permeability to this ion. While the potential difference is developed by an increased transfer of sodium ions (an electrogenic transfer of sodium ions) some anions must diffuse across the intestine to partially neutralize the transfer of charge. It is because they are delayed by the cell membranes that we can measure the potential difference and thus register the activity of the intestine's electrogenic pump.

### Sites of production of potentials

A number of other sites where ion transfer across epithelial cell membranes can be easily measured are shown in Fig. 45. This curious creature is a chimera of animal and man, hence its name, "animan"! It is designed to incorporate measurements obtained both from animals (mammals) and man. Electrical potential differences can be measured across the epithelial membranes of animan in the various places designated, i.e. its stomach, small intestine, large intestine, uterus and vagina. In each of these organs the epithelial cells lining the lumen are transferring ions across their membranes and generating electrical potentials because of the presumed net transfer of charge. The ion transfer picture however, of each organ is complex.

The stomach transfers hydrogen and chloride ions, it pumps both from the blood to form its acid secretion called gastric juice. The small intestine, as we have seen, transfers sodium ions slightly faster than chloride and other anions, hence the lumen is negatively charged. The large intestine not only secretes bicarbonate ions into its lumen but also reabsorbs sodium ions. Both these ion movements make the lumen negative compared to the blood. The cells of the uterus have only recently been shown to be involved in ion transfer that generates electrical potentials. Uterine cells (endometrium) appear to be able to pump both potassium and chloride ions into the lumen while they also probably absorb sodium into the bloodstream. The net effect is to make the lumen negative. The exact ion transfer causing the potential across the vaginal wall is

## THE ANIMAN

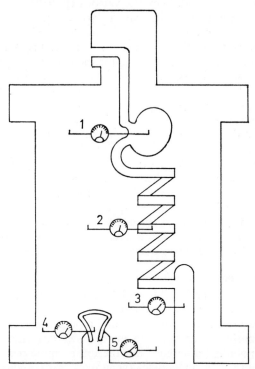

*Fig.* 45. Often when epithelial cells transfer ions from one body compartment to another they generate potential differences across the epithelial membrane. This strange creature, the animan, is designed as a composite of animals and man so that the electrical activity recorded from a variety of mammalian species can be used in one figure. The animan shows five regions where potential differences can be measured between the hollow lumen of an organ and the blood. The sites, potentials generated and ion transfers thought to be the cause of the electrical potentials are given below.

| Site | Organ | Potential (mV) (lumen negative to blood) | Ion transfer |
|------|-------|------------------------------------------|--------------|
| 1 | Stomach | 20–50 | $H^+$, $Cl^-$ ($Na^+$?) |
| 2 | Small intestine | 1–5 | $Na^+$ |
| 3 | Large intestine | 10–20 | $Na^+$, $HCO_3^-$ |
| 4 | Uterus (endometrium) | 10 | $K^+$, $Cl^-$, $Na^+$ (?) |
| 5 | Vagina | 20–60 | $Na^+$ (?) |

also not yet known. It appears that these cells transfer sodium from the lumen to the blood. Why the various epithelia of the hollow organs create negative potentials is not clear. The potentials, however, should be regarded merely as a "by-product" of the ion transfer. It is really the actual transfer mechanisms subserving absorption and digestion in the alimentary tract, and control of ionic environment in the reproductive tract (for sperm and ovum) that are the important processes. It is easy to overlook this when discussing potentials and to think erroneously that the job of these ion transferring epithelia is to manufacture electrical activity for the physiologist and biophysicist to measure and write books about! In fact the actual values of the potentials generated at each site may possibly be useful measures of the functional integrity or the performance of the various epithelia of the hollow organs. Thus when disease or malfunction strikes any of these organs, we may be able to measure its effects and assess the treatment for recovery simply by measuring the electrical behaviour across the different membranes. The hospital of the future may well be monitoring diseases of the alimentary and reproductive tracts electrically!

### Non-electrogenic ion pumps

1. **Ion exchange pumps.** Not all ion pumps in cell membranes are electrogenic. There are some that exchange one ion for another of the same charge, i.e. a sodium out and a potassium in. As there is no net transfer of charge, the pump itself does not create an electrical potential difference but the ion asymmetries developed by the pump may, of course, themselves bring about diffusion potentials. Another type of ion pump that is not electrogenic is to be found in the mucosal cells of the gall-bladder.

2. **Neutral ion pump.** Some epithelia transfer large amounts of sodium without generating a potential difference across their membranes. The reason is that the carrier that transfers the sodium also transfers an anion. Each action of the pump moves a sodium and a chloride ion at the same time, hence there is no unequal transfer of charge or current across the membrane. This type of neutral ion pump was first investigated in detail by Jared Diamond working in Cambridge in 1962. He found that the mucosal cells that line gall-bladders in a number of species possessed this type of ion pump. The pumps appear to have an absolute requirement for sodium and chloride ions, the carrier is unable to transfer any other type of ion across the cell membrane. If a gall-bladder is filled with

sodium sulphate solution no active transfer of sodium takes place from the lumen. If the solution is replaced with one of sodium chloride, the cells absorb both ions in a one for one ratio from the lumen. Replace the solution with potassium chloride and again no active transfer of potassium or chloride takes place. The mucosal cells of the gall-bladder thus appear to behave as if they have a site on the ion carrier that must be simultaneously filled with a sodium and a chloride ion before it can move across the membrane to transfer the molecule of sodium chloride to the other side.

### Experimental criteria for assessing active ion transfer in epithelia

The natural relationships between ion movements and electrical potential differences make the investigations of the mechanisms of ion transfer extremely complex. It is very difficult to assess whether an ion is being transferred by an active ion pump or whether it is moving passively under an electrical gradient. With ions we must know both the chemical and electrical gradients before we can say whether they are being concentrated by a pump or because of a favourable potential. While it is possible to measure both chemical and electrical forces across some membranes (nerve axons and some giant plant and animal cells) most membranes are more difficult to deal with, especially if they have a complex structure like that of frog skin. If the potential across the epithelium could be eliminated without damaging the structures then any ion concentrated in the bathing solution on either side of the membrane is likely to have become so by an active transfer mechanism. If we could literally short circuit the potential across the membrane ions would travel only according to their membrane permeabilities and their chemical concentration gradients – just like neutral molecules. We should thus be able to apply the same criteria for active transfer to them as to neutral substances (see page 99).

### The voltage clamp technique

The first experiments of changing the potential across a biological membrane to measure ion permeabilities and movements involved the squid giant axon once again. Two electrodes were placed across the axon membrane. One was pushed inside running parallel with the membrane, the other remained outside. By passing an external current between the two electrodes any desired potential could be imposed on the axon membrane. It could be increased (hyper-polarization) or decreased (depolarization). This technique was pioneered by American biologists in the early 1940s and was called the "voltage clamp technique".

**The voltage clamp technique applied to epithelial membranes.** The practical application of the voltage clamp technique to epithelial membranes was made by two Danish biologists in 1951. Hans Ussing and K. Zerahn used as their membrane a frog skin mounted as a sheet between two chambers. The skin, of course, had previously been removed from the frog! This arrangement where a tissue is kept alive by bathing it in oxygenated artificial salines away from

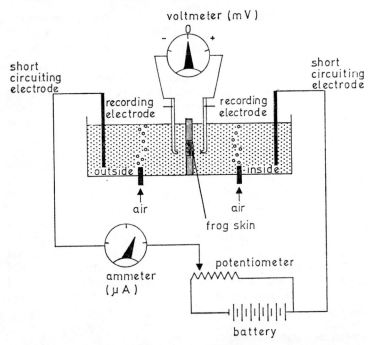

*Fig.* 46. Measuring the electrical potential difference across the frog skin which is mounted as a membrane separating identical bathing fluids gassed with air (in vitro preparation). The potential is sensed by the two recording electrodes placed close to the skin and the value is read from the voltmeter. In the diagram the skin is being short-circuited by application of an external current from a battery and potentiometer through the short-circuiting electrodes on either side of the skin at the end of the chambers. The external current needed to bring the potential across the skin to zero is called the short-circuit current. It is equal to the net sodium transfer across the skin (after Ussing and Zerahn 1951).

its normal habitat, is usually known as an *in vitro* experiment. The expression derives from the fact that in many experiments the isolated tissues were incubated in vessels made from glass, i.e. vitrum. The Ussing–Zerahn technique recorded the potential difference across the isolated frog skin by a pair of electrodes held close to the

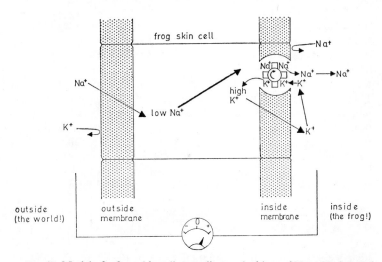

*Fig.* 47. Model of a frog-skin cell according to the ideas of Hans Ussing and Valborg Koefoed-Johnsen. The outer membrane is permselective to sodium while the inner one is permselective to potassium. A linked sodium–potassium ion pump is thought to be located at the inside membrane. The pump extrudes sodium from, and accumulates potassium into the cell. The net result is that sodium leaks into the cell down its concentration from the outside and is pumped into the frog's body in exchange for potassium. The potential across the skin is the sum of the diffusion potentials created (*a*) by sodium ions leaking into the cell at the outer membrane, and (*b*) by potassium ions leaking out of the cell via the inner membrane.

skin (Fig. 46). With another pair some distance away, a current was passed through the skin equal and opposite to that produced physiologically by the skin cells. The current for this outer circuit came from a battery and could be varied through a potentiometer, or resistance. By this means the potential across the skin could be brought to zero (as recorded by the electrodes near to the skin) by application of the external current. In such a situation they had the skin voltage clamped to zero and identical solutions bathing either side of the skin, thus the electrochemical potential of each ion was the same on both sides of the tissue and there could only be a net transfer of ions which were actively transferred, or pumped into one or other bathing solution. All other ion movements would occur by passive means, their net flux being zero.

In order to identify the movements of specific ions, radioactive tracers of sodium, and in other experiments, potassium and chloride, were used. The experiments revealed that sodium was the only ion that was pumped across the skin from the outside solution to the inside one. Later work by Ussing and Koefoed–Johnsen showed that the outside membrane was permselective to sodium (see page 111). These authors postulated a model of the frog-skin cell that had its outside membrane passively permeable to sodium but not potassium ions. The inside membrane, however, was found to be passively permeable to potassium but not sodium ions. Sodium had to be pumped out of the cells across this membrane by an active ion pump (Fig. 47). The frog-skin cell thus possessed two membranes with distinct and very different ion permeabilities.

### The "short circuit current" – the amount of current needed to clamp the frog skin to zero potential

One of the surprising features of the experiments by Ussing and his co-workers was that the actual value of the current needed to bring the skin potential to zero was practically identical with the net amount of sodium transferred to the inside compartment. This current was given the name "short-circuit current" as it was the amount of electricity needed to short circuit the skin potential to zero. As the short-circuit current equalled the net sodium flux all the current transferred across frog skin is carried by this ion. This was a major biological discovery and opened up the gates to a flood of studies on other important membranes. All types of epithelial membranes have been electrically short-circuited since; the stomachs of frog, rat and cat, a host of animal small intestines from silkworms to man, the large intestines from frogs and dogs, the corneas of rabbits and the bladders of turtles! In fact wherever a biological structure can be removed, stretched, mounted as a membrane between two chambers and kept alive *in vitro* for a reasonable time, a physiologist has had a crack at measuring short-circuit currents and ion fluxes. It should be noted, however, that while many epithelial membranes have been shown to have a close relation between net sodium flux and the short-circuit current, a number of tissues are known to have pumps for other ions. Thus these epithelia have all or part of their short-circuit current attributable to ions other than sodium.

Although the word "breakthrough" is now overused and hackneyed, the short-circuit current technique was a major breakthrough and revolutionized membrane physiology. It has allowed physiologists and pharmacologists to study with some degree of control, the events and complex details of the transfer of ions across cell membranes.

# CHAPTER 8

# Water Transfer

"Taint the things we don't know that makes us so ignorant,
it's the things we know that ain't so."

JOSH BILLINGS

Because the movement of water across membranes is so often linked
with osmosis and osmotic pressure, it is best to briefly examine what
these simple but loaded terms imply. Although the fundamental
concepts of osmotic phenomena were established many years ago
they are still, surprisingly, a target for dispute by different authors.
This highlights an important point – we still have not attained a
perfectly satisfactory kinetic theory of osmosis. The fault probably
lies in our poor understanding of the structure of water as a fluid.
It is only to be expected that we should have even less understanding
of the manner by which solutes affect this structure and influence
water movement. A number of facts, however, are absolutely
clear.

## Osmosis and osmotic pressure

When a solution is separated by a semipermeable membrane (one
that is permeable to water but not solute) from pure water, water
molecules pass through the membrane to dilute the solution. This
movement of water is known as osmosis. It may be stopped by
applying externally a pressure to the solution. If the pressure exactly
counterbalances the water movement into the solution, the pressure
that has to be applied is called the osmotic pressure. The nineteenth-
century botanists and chemists who investigated osmosis discovered
that the osmotic pressure corresponding to a given solution was
directly proportional to the concentration of solutes in the solution.
It was found to be equal to the pressure the solute molecules would
exert if they were in gaseous form occupying the same volume as the
volume of solution and was dependent on the absolute temperature.
Because solute molecules in solution appeared to behave as if they
were molecules of a gas, the idea developed that osmotic pressure

was due to the constant bombardment of solute molecules against the membrane. Some textbook authors even thought that these molecules blocked the passage of water through the pores so that the flux of water molecules from pure water to solution would be greater than that from solution to water! Modern studies on osmotic pressure and osmosis have shown up deficiencies in the early ideas and we now have better descriptions using a thermodynamic rather than kinetic approach.

## Modern concepts of osmotic flow of water

Water molecules are said to have a chemical potential such that water will always flow from a region of high chemical potential to one of lower chemical potential. Thus water flows down its chemical concentration gradient just like any other substance. We can alter the chemical potential of water either by applying an external pressure which increases its chemical potential, or by adding a solute which decreases its chemical potential. Using this new terminology, water flows into the solution through the semi-permeable membrane because the solute lowers the chemical potential of water in that compartment. Osmosis occurs then, because water moves down its chemical potential gradient.

So far this modern approach to osmosis has begged the question, how does the solute lower the chemical potential of water in a solution? One possible approach is to think that the solute lowers the actual concentration of water. Thus pure water has a higher water concentration than water in a solution, and the water molecules in essence move down their concentration gradient. This concept of a solute lowering the concentration of water can be easily visualized in a diagrammatic form. Consider a pore in a semipermeable membrane separating a solution from pure water (Fig. 48). As the membrane is semipermeable, by definition the pore will exclude any solute molecules so that it can only be filled with water molecules. Now the film of water and solute molecules at the end of the pore (film 1) will be opposed by a layer in the pore containing only pure water molecules (film 2). It is clear that the concentrations of water in this layer is greater than that in solution, there is thus much greater likelihood for water molecules to move from film 2 into film 1 than from film 1 into film 2. The net flux of water will thus be out of the pore rather than into it. This net flux would make the inside of the pore of lower density than that of the bulk pure water (designated in the diagram as film 3). Because of this, water molecules will move from the bulk solution into the pore, causing a hydrostatic pressure in the pore. The net result is that water moves down its concentration gradient from its high concentration in pure water to its lower one in solution.

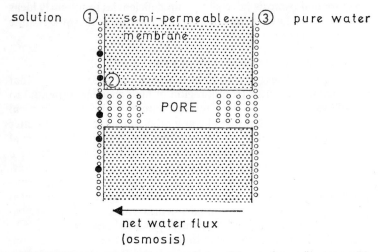

*Fig.* 48. Diagram representing a semipermeable membrane (i.e. permeable only to water) separating a solution containing a solute (black circles) from pure water (open circles). Film 1 represents a solution layer in contact with the pore, film 2 is that at the end of the pore and film 3 is that in contact with the membrane on the pure water side. It is clear that the concentration of water in 1 is less than that of 2 or 3. Water will thus move from 2 to 1, subsequently water will move from 3 to 2. The movement is called osmosis.

## Units of osmotic pressure

The normal units of osmotic pressure are millimetres of mercury or atmospheres. Biologists, however, have tended to use another unit of osmotic pressure called the "osmole". One osmole is the osmotic pressure of a 1M solution of an ideal solute (such a solute has a freezing-point depression of 1·86° C). Generally speaking the osmole is too big a unit for physiology so the milliosmole ($\frac{1}{1000}$ of an osmole) is used. With mammals, the osmolarity of their blood is approximately 300 milliosmoles/litre. Any fluid having an osmolarity above that of blood plasma is called hypertonic and anything below is called hypotonic. Solutions with the same tonicity as blood are called isotonic solutions. As mammalian cells are in osmotic equilibrium with the blood plasma, hypertonic bathing solutions will cause them to shrink in size while hypotonic solutions will cause them to swell and possibly even burst.

### The leaky membrane as opposed to the ideal semipermeable

Until now we have assumed that all membranes are ideally semipermeable, i.e. that the pores exclude solute molecules. We know that in the case of cell membranes they are far from ideally semipermeable for they allow the passage of electrolytes and non-electrolytes. Biological membranes would be better described as leaky membranes. The degree of their leakiness is often a factor of great influence in many physiological functions. Sometimes a body process in an animal is controlled by changes in the leakiness of the cell membranes. It is thus of some importance to be able to measure this property. We can do this by using a neat idea devised by A. J. Staverman in 1951. He proposed that for ideal semipermeable membranes the osmotic pressure measured across the membrane (observed osmotic pressure) should be identical with the theoretical osmotic pressure calculated for the particular concentration of solute used. The greater the leakiness of the membrane to the solute, the smaller would be the observed osmotic pressure. This is obvious, as a solute can only give rise to an effective osmotic pressure if the membrane constrains the solute in one compartment. If it is free to filter through the membrane its concentration gradient will rapidly dissipate by diffusion. Staverman assessed the leakiness of the membrane to a particular solute by dividing the observed osmotic pressure by the theoretical osmotic pressure; the latter is the pressure that would develop if the membrane were non-permeable to the solute. He called the fraction obtained the "reflection coefficient" of the membrane and gave it the symbol of the Greek letter sigma, $\sigma$.

Thus
$$\sigma = \frac{\text{observed osmotic pressure}}{\text{theoretical osmotic pressure}}$$

If the membrane was ideally semipermeable the observed and theoretical osmotic pressures would be equal and $\sigma = 1$. The more leaky the membrane to the solute, the smaller would be the observed osmotic pressure and the smaller would be the value of $\sigma$. If the membrane were completely permeable to the solute, $\sigma$ would be zero. This is because no effective osmotic pressure could be generated across the membrane if the solute penetrated the membrane as fast as water entered the solution. Thus the normal range of values that $\sigma$ could be expected to show would be from 0 up to a maximum of 1, i.e. characterizing a membrane completely permeable to a solute to one completely impermeable.

Staverman's reflection coefficient has been of great use to biologists as an easy method to quantitate a membrane's ability to discriminate between solutes of different sizes. For instance the mucosal cells of the small intestine have a $\sigma$ of 0·8 for urea (molecular radius 2·3Å) and one about 0.97 for lactose, the disaccharide of milk (molecular radius 5·4 Å). This indicates that the gut is fairly permeable to urea but is nearly impermeable to lactose. Presumably the size of channels that allow urea to pass through the membrane are between 4·6Å and 10·8Å in diameter.

## Pathways of water movement through membranes

Water molecules are in a constant state of flux across cell membranes. In fact the early idea of a dynamic equilibrium of chemicals between cells and their environment came from studies with water molecules labelled with deuterium, the heavy isotope of hydrogen. This labelling with $D_2O$ enabled fluid movements to be followed and was used as an index of the movement of ordinary water molecules.

Water molecules can in fact pass through a membrane in two ways. They can either diffuse or dissolve into the membrane's matrix, pass across and then leave to enter the aqueous bathing phase or they can pass through the membrane via water-filled pores.

The first type of water movement, that through the membrane matrix, is governed by the laws of diffusion. Each water molecule is a separate entity, uninfluenced by other water molecules. They cross the membrane at random, independent of one another. This type of water movement across a membrane is usually called "diffusive flow".

**Diffusive flow.** In diffusive flow there is no net flux of water across the membrane. It is measured by following the movements of tagged water molecules. Today we use radioactive tracers of either hydrogen (tritium $^3H$) or oxygen ($^{18}O$) to label water rather than the non-radioactive deuterium. The tracer is added to the fluid bathing the membrane and either its appearance in the intracellular compartment or its loss from the bathing solution is used as a measure of the membrane's permeability to water. The diffusive permeability as measured by this technique is quoted in the same units used for other solutes, namely flux per unit area per second, i.e. cm/sec (see page 90). In reality what we are measuring is the unidirectional flux of water across a membrane where the driving force is only random thermal movement (diffusion).

**Bulk flow or osmotically induced flow.** If we applied an osmotic (or hydrostatic) pressure difference across a porous membrane there would be a net flow of water. By increasing the solute concentration on one side of a membrane, we would cause the water molecules in the pore to move down their concentration gradient into the solution. These would of course be followed by further solvent molecules. As they move through the pore they would entrain others in their stream and sweep them through. Solvent molecules would thus have a "solvent drag" effect on their own species (see page 78). In this situation a net transfer of fluid is taking place, quite different from diffusional flow where there is no net flux. We measure a unidirectional flux powered by a concentration gradient for water. We should not overlook the fact that even when an osmotic pressure difference is used to induce bulk flow through pores, diffusional flow through the membrane matrix can still occur. In fact if the membrane has no pores, an osmotic gradient could only cause water movement by diffusion. The water would dissolve in the membrane matrix, pass across and diffuse into the solution down its activity gradient. Some cells have membranes (the giant marine alga Valonia is one) that appear to have no aqueous pores. In this case water movement during osmosis is by diffusion only.

Osmotic permeability of cells to water is usually estimated by measuring the rate of swelling or shrinkage of cells and equating these volume changes to changes in water movement into or out of the cells. With some tissues (especially epithelial ones) it is possible to directly measure the actual amount of water transferred across the cell membranes.

Many units have been used as a measure of osmotic permeability of water; one that has been the most often used in the past is the amount of water in cubic microns ($\mu^3$) passing through each square micron of cell membrane ($\mu^2$) in one minute per atmosphere of pressure difference across the membrane, i.e. $\mu^3/\mu^2/\text{min/atms}$. Unfortunately many other units of osmotic permeability have been used by different workers which leads to much confusion. The data in Table 13 were compiled by David Dick and shows the

F

complicated variations in units that can be wrung from this simple measurement.

TABLE 13. *Various units of osmotic water permeability (after Dick 1966)*

| |
|---|
| 1 cm/sec = $10^4$ $\mu$/sec |
| 449 $\mu^3/\mu^2$ min atm 20° C |
| $7.23 \times 10^{-7}$ cm$^3$/cm$^2$ sec cm H$_2$O pressure at 20° C |
| 18 cm$^3$/cm$^2$ sec osm per cm$^3$ |

## Differences between membrane water permeability measured by diffusional flow and by osmotic induced flow

A number of experiments using different cell types showed quite clearly that when the permeability of membranes to water was measured by using diffusion as the moving force for the tracer molecule, the result was very different from permeability measurements obtained when osmotic induced flow was created. The membrane permeability to diffusional movement of water always appears to be less than that measured in the osmotic experiments. Examples of the difference between the two types of permeabilities have been found for cells as varied as squid axons and human red blood cells. In the former, osmotic permeability is 49 $\mu$/cm, while diffusional permeability is 1.3 $\mu$/cm, in the latter cells the figures are 127 $\mu$/cm and 53 $\mu$/cm respectively.

A number of scientists took this difference to indicate that in osmotic flow water must move in large packets or in bulk. To allow this to happen there must be channels in the membrane that allow bulk water movement. It was argued therefore, that the difference between the diffusional and osmotic permeabilities for water indicated that cells had aqueous channels or pores. For some time this difference between permeabilities was regarded as strong evidence in favour of pores in cell membranes. Unfortunately a simple factor was overlooked that made this argument very much weaker; this was the fact that solid surfaces affect the character of water.

## Water in contact with surfaces – the unstirred layers

Although physical chemists have known for some time that water in contact with a surface becomes a structured layer with a much higher degree of organization than the bulk fluid, the importance that this structuring of water had on the permeability of cell membranes was not realized until quite recently. It came about because Jack Dainty, working with plant cells in Edinburgh University, found great difficulty in obtaining accurate measurements of the diffusional water permeability in these cells. He began to realize that part of the difficulty was because he had not taken into account that the layer of structured water close to the membrane of the cell constituted a "membrane" of orientated water molecules. Tracer molecules of water used for measurements of the diffusional permeability had not only to penetrate the cell membrane proper but they also had to get through this layer of structured water. In reality there were two membranes in series. The structured water layers are called "unstirred layers" and they can be surprisingly thick, even when the solution bathing the cells is highly agitated. Even when there is violent stirring, the surface of the cell membrane imposes an order on water molecules as far as 10 $\mu$ from its surface. When cells are amassed in tissues, solutions in contact with their membranes are usually not well mixed. Under these conditions unstirred layers could be some 500 $\mu$ thick. The resistance of a layer of water this thick to diffusion of water molecules is in some cases more effective a barrier than a cell membrane! Clearly a correction for the unstirred layers has to be made if we want the real value of the permeability of water through the membrane. This point was overlooked until the unstirred layer concept was brought to biologists' attention by Dainty and his co-workers about 1963. In many cases it is the unstirred layers that are the major cause of the diffusional permeability being less than the osmotic permeability. At first it appears that the unstirred layers should affect the measurements of both osmotic and diffusional permeabilities to the same extent. But this is not so. In the experiments where an osmotic difference is used to cause the movement of water, the induced water flow through the membrane has a finite velocity and sweeps away part of the unstirred layer. Thus the unstirred layers have only a small effect on osmotic permeability; it is only about 2 per cent less than the theoretical value. With diffusional permeability, however, the flux of water through a 10 $\mu$ layer of water is in fact equal to the permeability of many cell membranes. This is because most cell membranes are so permeable to water that they behave practically as if they were a layer of water!

Although unstirred layers are important it should be realized that the less permeable a membrane is to water, the less effect the unstirred layers will have on the measurements of diffusional permeability. Obviously if a membrane is 100 times less permeable to water than an unstirred layer of equivalent thickness, most of the resistance to the passage of water across this system will come from the membrane. While the presence of unstirred layers has caused us to reconsider the use of the difference between permeabilities as evidence of pores in cell membranes, this does not mean that the idea of their presence has to be relinquished. There are, of course, other pieces of evidence for these tantalizing structures, as we have seen in the earlier historical chapters of this book. We should, perhaps, just recollect what these were and re-examine our ideas of pores in cell membranes.

## Pores in cell membranes

The original idea of pores in cell membranes is an old one deduced from experiments in which the movements of various substances were used to probe the physiological properties of membranes. These experiments and their conclusions have been previously related in Chapters 2 and 3. The dimensions of pores have been probed by using solutes of different shapes, charge, molecular volume, lipid solubility, etc., and assessing their penetration or nonpenetration. The data obtained gave rise to the concept of the ideal circular pore, sometimes called equivalent pore. This "equivalent pore" is in fact a hypothetical one manufactured to explain the results obtained in the laboratory. If the data of an experiment fits in with the concept that the membrane behaves as if it had pores set exactly at right angles to the bathing solutions, with a constant diameter right the way through the membrane, then it is stated that the results are consistant with the membrane having right, circular pores of so many Ångstroms in diameter.

Table 14 contains a collection of equivalent pore radii measured in a number of cells by a variety of techniques. It is surprising how close to a pore of radius 4Å these measurements approximate. Estimates of the total surface area of the cell membrane given over to pores varies widely, depending on the cell's functions and the method of estimating the pore area. There is one common factor, however, in all the estimates – the pore area is always a very small fraction of the total membrane. Red cells are thought to have between 0·01–1 per cent of the total membrane as pores, capillaries in mammalian muscle have 0·1 per cent while the highest estimate is in stomach mucosa of the mouse where some 3 per cent of the total membrane is "pore area". The experimental difficulties in making such measurements and calculations suggests that these figures should be treated as estimates only. No one knows, even today, how near these ideal hypothetical pores match up to real structures in cell membranes, that is if they exist at all! Perhaps the best current description of pores in biological membranes comes from a group

TABLE 14. *The equivalent pore radii of cell membranes and the crystal radii of some important ions. The radii of hydrated ions depend on the number of shells of water molecules that are measured. Different methods of measuring the degree of hydration give different radii. There is thus no single figure that can be quoted for the hydrated radii. Current opinion is that the hydrated sodium ion is larger than the hydrated potassium or chloride ion*

| Cells | Equivalent pore radii (Å) |
|---|---|
| Human red blood cell | 4·2 |
| Squid axon | 4·3 |
| Mucosal cell (small intestine) | 4·0 |
| Toad skin (outer membrane) | 4·5 |
| (inner membrane) | 7·0 |
| Frog egg | 2·1 |
| Dog red blood cell | 7·4 |
| Amoeba proteus | 6·9 |

| Molecules and ions | Crystal radius (Å) | Hydrated ion radii in solution |
|---|---|---|
| Water | 1·4 | |
| Sodium | 0·97 | $Na^+ > K^+ \simeq Cl^-$ |
| Potassium | 1·33 | |
| Chloride | 1·81 | |

of South American biologists who have been studying membrane permeability. They wrote "the expression equivalent pore radius should be taken in its literal meaning, since at the present time no experimental evidence is available concerning the conformation of the pathways . . . they could be approximately cylindrical, rectangular or intermediately shaped, tortuous or straight, and they may exist as transient paths appearing and disappearing by random fluctuations or as permanent structures. They may be lined by lipid polar groups or by proteins, or may be formed by spaces separating the outer surfaces of membrane repeating units. They may exist or occur individually at intervals, or in a small cluster."

While we have examined in some detail the arguments for and against pores with regard to permeation, both of solutes and of water, it is profitable to mention two new experimental approaches that give results consistant with a porous cell membrane theory.

### Pores in epithelial cells

**Streaming potentials.** The possible existence of pores has been highlighted in an unusual manner by using an osmotic flow of water through a membrane to generate electrical potentials. This phenomenon at first sounds rather curious and ridiculous, how can water flowing through a membrane generate electrical potentials? The answer lies in the structure of the pores in the membrane. If they are charged like those shown in Fig. 41 they will contain a large preponderance of ions of a charge opposite to that of the pore walls. Negative pores will be filled mainly with positive ions and vice versa. If we now cause a stream of water to move through these pores then the water molecules will flush out some of the ions contained in the pores. Hence the water stream will carry or entrain more of one type of ion than another to the compartment to which the water is flowing. This compartment will thus become electrically charged. Water movement generates a small net transfer of charge across the membrane by solvent drag. It does not matter whether the water is caused to move by an osmotic gradient or by hydrostatic pressure; a potential will be generated by the water streaming through the charged pores of the membrane. This potential is known as a streaming potential or an electro-kinetic potential. Any tube through which a liquid is flowing will generate these potentials and it is fairly easy to create such streaming potentials in the laboratory with small glass tubes or artificial membranes. With biological membranes only two candidates for this sort of potential have so far appeared. One is the epithelial cell of the gall-bladder and the other is the mucosal cell of the small intestine. If we pull water through the mucous membrane of either of these tissues by applying an osmotic gradient, we can generate quite large potentials across the membrane. An osmotic gradient of 200 milliosmoles can generate some 10–15 millivolts.

These streaming potentials are interesting for another reason; their polarity should indicate the charge of the pore walls. If the pores are negatively charged, pulling through fluid should make the side to which the fluid is moving more positive (as positive ions rather than negative ions are swept through). With both intestine and gall-bladder the compartment to which the fluid is moving actually does become positively charged. This must mean that the structures in the mucosal cells through which the fluid passes are negatively charged, a result that confirms the experiments where diffusion of a salt across the small intestine also indicated a negative pore polarity (page 117). One word of caution, however, is needed before we rush off and use streaming potentials as a weapon to hammer into submission those who doubt the existence of pores in the cell membrane. Although the interpretation of the streaming potential experiments indicates a pore structure it does not tell us where this structure is located in the sheets of mucosal cells. While it may be in the cell membranes it is also possible for the pores to exist between the cells. As yet we have no way of proving that the latter is a wrong interpretation.

### Pores in nerve membranes

When a nerve is at rest the inside has a high potassium and a low sodium concentration compared to the low potassium and high sodium concentration in the bathing fluids. During the passage of a nerve impulse the membrane becomes altered so that the external sodium ions flow through the membrane into the axoplasm. A few milliseconds later, the membrane's permeability changes and allows the internal potassium ions to pass out down their concentration gradient. As these minute passive movements of sodium and potassium occur with every impulse the ionic concentrations of the axoplasm would change unless ion pumps in the membrane pushed out the sodium and pumped back in the potassium resetting the cell for further conduction. Our focus, however, is on the passive movements of ions, for these must flow through the membrane in some sort of channel or pore. It is only recently that it has been possible to directly estimate their number and spacing in the axon membrane. There is strong evidence indicating that two types of channels exist, one set for potassium diffusion out, the other for sodium diffusion in. While the experiments detailing the number of sodium pores in axon membranes were executed only a short time ago, the story of the chemical used for their identification stretches back well into the late eighteenth century.

**Tetrodotoxin – a specific blocking agent for sodium pores.** The seas around Japan and the Pacific islands abound with a fish species called puffer or globe fish that are thought highly of as a delicacy. The fish are known, however, to manufacture a highly toxic substance and care has to be taken in their preparation. There is an entry in the diary of the explorer Captain Cook telling us just how poisonous they are. His ship, the *Repulse*, had stopped at some Pacific Islands, in September 1774, where two of his naturalists were given some fish by the friendly natives. Because the ship's cook had little time for their preparation he gave the Captain and his friends only their livers and roe, throwing the intestines, etc., to a pig kept on board. The following morning Cook and the scientists had numb mouths and tongues and had lost the full strength of their limbs. Fortunately they recovered but the pig died! Many people in Japan still die each year from the poison of puffer fish, in 1957 there were some 90 deaths out of 175 reported cases of puffer fish poisoning. Japanese scientists have, of course, been interested in the toxin for many years and they succeeded in making a crude extract and named it tetrodotoxin. In 1966 both Japanese and American research workers finally isolated the pure substance and defined its unusual nature ($C_{11}H_{17}N_3O_8$). It is one of the most toxic non-protein animal poisons known.

We now know that it kills because it prevents the nerve axon from conducting nerve impulses; specifically, it blocks the passive entry of sodium into the axoplasm. It appears to have no other action and is only effective when applied to the external face of the axon membrane. Because of these properties it is a highly specific tag for the sodium pores. This property provides an unusual way of estimating their number. We can use the toxin as a label to count pore sites, just like carrier sites have been counted by measuring the uptake of specific inhibitors (see page 105). On the reasonable assumption that one molecule of toxin blocks one sodium channel, it was found that some 13 molecules of the toxin were bound per square micron ($\mu^2$) of nerve. Thus there can be at the most only 13 sodium pores per square micron of nerve fibre. Other calculations indicate that these channels for sodium would be sited some 800 Å or $0 \cdot 1$ $\mu$ apart. It is thus clear that nerve axon membranes have very few "pores" per unit area and that any chance of seeing them by electron microscopy is unlikely.

## Mechanisms of water transfer across biological membranes

While water is in constant diffusional movement across all cell membranes, some cells are specialized to transport water from one compartment to another. In these cases, the water is moved across two cell membranes and the movement is called transcellular water transfer. Many epithelia undertake transcellular water movement and either secrete or absorb large quantities of what is normally an isotonic fluid. For instance the salivary glands secrete saliva, the stomach gastric juice, the liver bile and the pancreas pancreatic juice. In a human these glands pour into the lumen of the small intestine at least 7 litres of fluid in twenty-four hours. This volume is normally completely reabsorbed by both small and large intestines.

What are the mechanisms that cause the transcellular movement of fluid?

### Transcellular fluid movement

A short list of the possible mechanisms that give rise to fluid transfer across an epithelium would include:

1. Filtration.
2. Classical osmosis.
3. Electro-osmosis.
4. Pinocytosis.
5. Codiffusion.
6. Local osmosis – specific variations of classical osmosis based on localized concentration of solute.

None of these mechanisms are exclusive, any or all could be operating at the same time. Some of these mechanisms were examined in general terms in Chapter 5. Let us specifically consider each one in relation to transcellular water transfer.

**1. Filtration.** We have already discussed the movement of fluid by filtration in Chapter 5. Obviously this only occurs in biological systems that can generate a hydrostatic pressure. Examples of fluid moving across membranes as a result of hydrostatic pressure are the formation of tissue fluid from blood and the formation of Bowmans capsule fluid in the kidney in mammals.

Hydrostatic pressure and filtration play an insignificant role in the absorption of water from the intestine or in the formation of secretions by various glands. For one thing many glands can secrete against a pressure much higher than arterial blood pressure, thus they actually do work to produce their secretions and appear not to use the hydrostatic pressure developed by the heart.

**2. Classical osmosis.** Water will move across biological membranes by classical osmosis wherever there is a difference in solute concentrations in the fluids bathing the membrane. This difference in concentration could come about by active pumps transferring either ions or non-electrolytes across a membrane. As we know that cells have active pumps for ions and non-electrolytes, it has often been suggested that osmosis is likely to be the major mechanism of transcellular fluid movement, the cell producing a high concentration of solute which then causes water to move from one compartment to another by osmosis. Under these circumstances water transfer is passive or secondary to net solute transfer, without solute transfer there can be no fluid transfer.

One of the practical difficulties in accepting this explanation of fluid transfer has been with epithelia that transfer large quantities of fluid across their membranes from one compartment to another. More often than not the fluid transferred has an osmotic pressure or osmolarity identical to that of blood or plasma. In some situations it has been found that these epithelia (the small intestine is the one commonly used) actually appear to transfer water from a low chemical potential to a higher one. That is, the intestine apparently can undertake active transfer of water! This cannot, of course, occur thermodynamically unless the cells have an active water pump able to transfer pure water molecules across the membrane. Most biologists are loath to accept such a pump and prefer to keep the idea that without net solute transfer there can be no net water transfer. But how then to explain this uphill transfer of water by the intestine? A possible way out of the dilemma was found by Jared Diamond in his experiments on the absorption of fluid across the gall-bladder. He showed that there was no need to postulate active pumps for water. Osmosis was still the cause of the water movement but the solute transfer that caused the osmotic flow occurred into a local tissue compartment removed from the fluids bathing the epithelium on either side. Diamond found that, in fact, the structure of gall-bladder epithelium formed by the sheet of epithelial cells had properties different to the individual cells of which it was composed. We shall investigate these properties after we have dealt with the next two methods of fluid transfer.

3. **Electro-osmosis.** When an electrical potential difference is imposed across a fixed porous membrane by two electrodes, water moves through the membrane under the influence of this potential. The movement of water is called electro-osmosis. It is the converse of streaming potentials, where moving fluid through a charged membrane causes a potential difference. The movement of water through the membrane (because of the potential) is due to the fact

that the surface of the pores have a fixed charge; that the water in contact with them is charged and that the pores contain charged ions. When the potential is applied the fixed membrane cannot move but the water and ions do. Electro-osmosis can be easily demonstrated in the laboratory using inert, porous membranes and current supplied by a battery. Does such a mechanism operate to transfer fluid across cell membranes? This was a question first asked nearly 100 years ago and it is only lately that an answer has been given. As cell membranes often have a potential difference of some 90 mV or less across their membranes it might at first be thought that the answer would be in the affirmative. In fact this magnitude of potential cannot drive a significant fluid movement. One needs potentials in the order of 1 volt (1000 mV) or more in order to obtain significant movement of water across cell membranes. Electro-osmosis could not account for the very large amounts of fluid moved across absorbing or secreting epithelia.

4. **Pinocytosis.** This mechanism of fluid uptake has been dealt with in detail in Chapter 5. While many free cells, like amoebae, use this method of absorbing water, it is not thought to play a major part in the transfer of large volumes of fluid during secretion or absorption by epithelial cell sheets. It was calculated by Dennis Parsons at Oxford University that if the small intestine absorbed its twenty-four-hour fluid load (i.e. 7 litres) by pinocytosis alone, then its cells would be required to make pinocytotic vesicles at the rate of 1,000 a second for the whole twenty-four hours. As this rate of vesiculation is some ten times that normally observed, it is highly unlikely that such a process alone could transfer the vast amount of fluid handled by the alimentary tract.

5. **Codiffusion.** When a molecule of a solute diffuses through a membrane via aqueous pores the solute is usually accompanied by a few hundred water molecules. This process is called codiffusion and is probably due to the frictional drag that the solute molecules exert on the water. It is an example of solute–solvent interaction. Co-diffusion can actually cause water to move against its own concentration gradient, that is, uphill. A simple laboratory experiment dramatically illustrates this point. A collodion membrane separates two solutions. One contains a permeant solute in high concentration while the other contains an impermeant solute in low concentration. Now the water activity is low in the compartment containing the high concentration of permeant solute and high in the one containing the low concentration of impermeant solute. Water should thus

move by osmosis from the high activity to the low activity compartment, that is from the solution with the impermeant to the solution with the permeant solute. In fact this does not happen. The permeant solute, because it flows through the membrane pores down its concentration gradient takes with it many hundreds of water molecules which counterbalance and even exceeds the normal osmotic flow. Hence there is an anomalous movement of water, apparently up its concentration gradient.

While codiffusion is a possible way for water to be transferred across some biological membranes, by the diffusion of electrolyte or non-electrolyte molecules down their concentration gradients, it does not explain the way that epithelia (such as the gall-bladder) transfer a solution that has an osmolarity practically identical with the solution that bathes the mucosal membranes of the epithelial cells. Other explanations have to be put forward for this phenomenon.

6. **Local osmosis – solute transfer into a confined space.** The story of local osmosis began in 1962 when Peter Curran and John Macintosh of the Biophysical Laboratory at the Harvard Medical School in Boston, Massachusetts published an ingenious artificial model system for biological water transport. The experimental model is illustrated in Fig. 49. It consisted of a plastic chamber separated into three compartments by two membranes. The one between compartments 1 and 2 was cellophane (membrane A) and that between 2 and 3 was a sintered glass disc (membrane B). During experiments, compartments 1 and 3 were open but 2 was closed. A graduated pipette was sealed into 3 to measure the flow of water into this compartment. Any increase in flow of water into 3 must be due to water moving in from 1 through 2 for 2 was a fixed, closed volume. When water or identical solutions were placed in all three compartments no volume changes of any significance were recorded. When, however, solutions of sucrose were placed into 1 and 2 while 3 was filled with water, fluid always flowed into 3 as long as the concentration of the sucrose in compartment 2 was greater than that of 1. Now if this movement occurred when 1 contained a solution of sucrose and 3 contained pure water, the water apparently moved from 1 to 3 uphill against its concentration gradient! As we have stated before this is thermodynamically impossible unless the membrane does work and can pump the water uphill. As yet inert glass or plastic membranes have not been found to pump water. What then is the explanation of the net water transfer? The answer

lies in the osmotic and hydrostatic pressures across the various membranes. In the system we have two membranes with separate reflection coefficients for sucrose, one for membrane A and one for membrane B. The cellophane membrane has a relatively low permeability ($\sigma = 0.4$) for sucrose while the sintered glass disc, which has very large pores, is nearly completely permeable to sucrose and

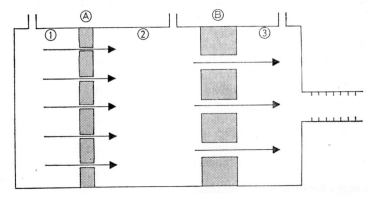

*Fig.* 49. A working model for water transfer by cells. A plastic chamber is divided into three compartments 1, 2 and 3 by the cellophane membrane at A and the porous glass disc at B. The compartments can be filled with solutions of different osmotic pressures by means of the inlet tubes. The tubes in 2 and 3 are usually sealed after filling so that 2 is a completely closed compartment of fixed volume. Movement of fluid into chamber 3 is indicated by the graduated tube attached. Fluid is transferred from 1 to 3 whenever chamber 2 contains a solution of solute (sucrose) of higher concentration than that in chamber 1. The model is analogous to an intestinal mucosal cell, A is one membrane (mucosal) and B another (serosal) while chamber 2 represents the inside of the cell (after Curran and MacIntosh 1962).

thus has its $\sigma = 0$. Therefore sucrose can give rise to an appreciable osmosis across the cellophane, but not across the sintered glass disc. If the sucrose concentration in 2 is higher than 1, water will move into the central compartment from 1 but not from 3 because there is no effective osmotic pressure generated across membrane B. The entrance of water into 2 will generate a hydrostatic pressure as the compartment is closed. Because the small pores in the cellophane A offer a higher resistance to water flow than the large ones in the sintered glass disc B, fluid will take the easiest path and move into 3. The net result is a transfer of water from compartment 1 to 3 even though 1 has a lower water chemical potential than 3.

Peter Curran suggested that water transfer by the small intestine might take place by a cellular mechanism of a type similar to this artificial model. In this case the concentration gradient of solute between compartment 1 (the solution bathing the luminal membrane) and 2 (the mucosal cells) would be maintained by active transfer of solutes: the major solute actively transferred being sodium ions, although other ions and even non-electrolytes would have their osmotic role. Water would follow the solutes by osmosis through the cell membrane. It was intuitively hinted that membrane A could possibly be the luminal membrane of a mucosal cell while B might be its serosal membrane (see Fig. 49). There was, however, no proof that either membrane could be so identified.

The artificial model created great excitement and explained the apparent active transfer of water by the intestine. It was the current explanation of fluid transfer until quite recently when Jared Diamond began a detailed study of the epithelium of the gall-bladder and found an intimate connection between changes in its ultra-structure and fluid transfer.

### Local osmosis in the gall-bladder

The function of the gall-bladder epithelium is to absorb fluid from the bile that the liver secretes continuously. By concentrating it some ten times and storing it, a strong solution of bile salts can be squirted into the duodenum at the appropriate time during feeding, to aid in the digestion of fats. In concentrating bile, the gall-bladder epithelial cells can absorb a remarkable amount of fluid, something like 10 ml/hr/gm. They do this by actively transferring sodium chloride molecules by their neutral pumps (see page 130). This transfer creates a high solute concentration and water follows passively by osmosis (Fig. 50). Yet the strange thing is that when one measures the tonicity of the fluid transferred it has the same value as the blood, i.e. it is isotonic with plasma. The compartment into which the sodium chloride is pumped must thus be, to a large extent, physically removed from the solution bathing the "back-doors" or blood side of the cells, otherwise we would expect to measure at least a small difference in osmotic activity between the two solutions. When the layers of epithelial cells were examined during fluid absorption and in conditions where no fluid was absorbed (poisoned cells) Diamond noticed that the distances between adjacent cells were very different. During fluid transfer the spaces between cells were distended and were about 1 $\mu$ thick. When

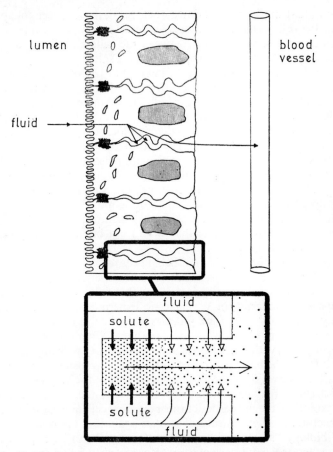

*Fig.* 50. The transfer of fluid by an epithelial sheet of cells. The cells (gall-bladder) are joined together to form a sheet by their "tight junctions" (black areas). The long, tortuous channels formed between cells (lateral spaces) gradually increase in size until they open out to the underlying connective tissue. The cells actively pump solute (NaCl) into the lateral spaces and the water flows from the lumen through the cells into the spaces by osmosis. The inset diagram of a lateral space shows how the ion pumps manufacture a high local concentration of solute, probably in the upper part of the lateral space. The fluid moving into the space by osmosis gradually dilutes this solute so that by the time it leaves it is fully equilibrated and is isotonic. It is then absorbed into the blood stream. The mechanism of fluid transfer by localization of an osmotic gradient in the lateral space has been called "standing gradient osmotic flow" (after Diamond and McD. Tormey 1966).

the water transfer was cut down or stopped, these spaces collapsed and the cells came very close together. The simplest explanation was that water was flowing between the cells and distending the lateral spaces. It was thus proposed that the solute pumps for sodium and chloride were located at the lateral borders of the cells, and it was here that the high concentration of sodium chloride was maintained. Water would diffuse through the lateral membrane, as both salt and water moved down the tortuous path between the lateral membranes, gradually diluting the high local concentration of salt, until by the time the fluid emerged at the free, open end of the spaces between cells the fluid was isotonic with plasma. In other words it has equilibrated with water to become isotonic, i.e. one molecule of NaCl is accompanied by 370 molecules of water. The solute pumps, by pumping the sodium chloride into the confined spaces between the lateral membranes, could generate a constant standing osmotic gradient of solute that would not be immediately swept away by the fluid stream.

As yet this concept that epithelial cells, *en masse*, have an ultra-structure that allows local osmosis to be the effective driving force for secretion or absorption is so novel that few critical experiments have been undertaken. We have no experimental proof that the neutral sodium pumps are indeed on the lateral membranes. It is at the moment only a working hypothesis. We must await further experiments with other tissues before we can say with any certainty that this is the basic process of isotonic fluid transfer. It does illustrate, however, the concept first mentioned in the very early pages of the book (page 3) that the properties of epithelia are sometimes greater than the properties of the individual cell units forming their structure. It also illustrates that once again that old hoary dogma of biology "structure is related to function" can be applied one stage lower, i.e. ultrastructure is also related to function.

# CHAPTER 9

## Looking Through a Glass Darkly

"After that, Mr Smith, I am none the wiser!"
"No, m'Lud, but your Ludship is better informed."
ALEX COMFORT

An American Committee of distinguished scientists was asked in 1937 to predict the future thirty years on. Their predictions missed radar, nuclear fission, computers, jets, rockets and transistors. They were right, however, when they said that aeroplanes would get bigger, safer and more comfortable but they blotted even this copy-book answer by saying they would not get faster! The rapidity of technological invention and the explosion in scientific education has made soothsaying a chancy business. In fact the title of this final chapter "Looking Through a Glass Darkly" was chosen to indicate the difficulty a biologist has in trying to make forecasts on the possible developments of his particular branch of biology. It is so difficult to be an accurate prophet in scientific fields that it would appear foolish to attempt the task. Yet, having said all this, one is still tempted beyond control to forecast. Scientists must be imbued with a naïve curiosity that compels them to stare ahead at the murkiness of the future and make definable shapes, or perhaps the answer is simply that the pride in making a correct forecast outweighs the embarrassment of being wrong! Forecasting future biology is fraught with difficulties, because many of its advances will rest, to a large extent, on advances in the exact sciences of physics, chemistry and mathematics. Some qualification is needed, however, about the difficulty of forecasting. It is not hard to see where a particular line of study is going to lead to for a few weeks or months ahead. The real difficulty comes when we try to forecast many years ahead, say for the year 2000. This is because the significant advances in sciences are usually of "ground breaking" dimensions. One writer said that scientific research is like a huge and very heavy flywheel. Normally it rotates very slowly, moved by the countless numbers of small, modest advances. These are the hack or routine experiments that go on in all sciences. Every now and again, however, a new brilliant idea or experiment jerks the wheel into a faster motion. When this occurs it takes everyone else with it by its momentum. The jerk is that overworked word again – the breakthrough. The momentum of the breakthrough is gradually dissipated, however, and the wheel turns again slowly. The energy in that field of investigation runs down until the next exciting impetus. The time between these can be many months or even years. The analogy of the wheel has its cruel side for its carries the implication that most research must need be of medium or even mediocre quality. This raises the question of why do we support it? A simple answer is perhaps that we know of no other way. One can make out a strong case that breakthroughs in biological sciences usually come from men who have had vigorous apprenticeships in their subject. Their early studies may have shown only medium promise but they were laying the foundations for the classic observation or the critical set of experiments to come. Obviously every biological scientist cannot win a Nobel prize but each can play his role in farming any newly discovered scientific virgin territory. Who knows, a piece of work from one of these honourable routine investigations might well be the spark that sets alight the touchpaper of a more inventive brain. The Nobel laureates may be the Generals of our sciences but even Generals need their armies of lesser men to help score their victories!

Interesting though this line of thought may be, it is taking us away from the main idea of forecasting the future of membrane studies.

### Where do we go from here? Or attempts at inventing the future!

Although there are many areas of current research that may show profitable results and increase our understanding of cell membranes there are three outstanding ones that could well yield important findings.

1. **Methods by which membranes are synthesized.** Without doubt our ideas of the structure of cell membranes are at a crucial point. Are they a bilipid layer with the protein stuck on the outside (i.e. the paucimolecular model Fig. 7) or are they made up from hosts of repeating units, each unit having a protein backbone with the lipid hung on or adsorbed on to it, as in the models of Green and Gross (Figs. 18 and 29)? Edward Korn, of the American National Institute

of Health Laboratories has recently suggested a method by which we could distinguish between these. His idea, like all good ideas, has that essence of simplicity that makes it satisfying. He proposes that if the membrane is the bilayered paucimolecular structure then it would probably be manufactured in the cell by synthesizing first the bilayer and then adding on the protein coats. On the other hand if the membrane is a construction of repeating sub-units with their protein backbone then it is likely that their sequence of synthesis would be reversed. The protein backbone or core would be made first and then the lipid would be attached or hung on to this core. Careful investigations of cellular membrane synthesis may tell us which of these pathways is correct. Of course we might well find that the membrane is a mixture of both types and that a definitive answer one way or another cannot be obtained. But, as is so often the case the cry must be "Don't think, do!"

2. **Hormones and their control of cell permeability.** The second area that merits careful study is how changes in membrane permeability affect cell function. We know that there are a number of hormones that change the permeability of cell membranes. Insulin, secreted from the endocrine portions of the pancreas, facilitates the passage of glucose across muscle cell membranes, while antidiuretic hormone (ADH), the chemical messenger secreted by the posterior pituitary, changes the permeability of the kidney tubules to water facilitating the reabsorption of tubular fluid. In both these cases a hormone affects membrane permeability and controls a fundamental physiological mechanism by this process. Other hormones have membrane actions but they have not yet been so clearly defined. For example oestrogens, the female sex hormones, are known to affect the entrance of substances like amino acids and ions into smooth muscle cells. Some hormones are now thought to control permeability by attaching themselves to the cell membrane and then causing the release of a second internal (i.e. intracellular) chemical messenger that changes membrane permeability and alters metabolic pathways by activating or depressing enzymes. The experiments to prove or disprove this new idea of a primary hormone, the first messenger, liberating a second intracellular messenger that actually accomplishes the task of changing membrane permeability are still being undertaken in research laboratories all over the world. New exciting ideas are being evolved and the structure and function of cell membranes obviously becomes infinitely more complex than that of a simple permeability barrier formed by lipid.

3. **Defects in transfer mechanisms in various genetic diseases.** The third line of research that will bear fruit is the study of animals or living organisms known to suffer from an absence of a cellular transport mechanism. For example occasionally humans have a congenital absence of particular active transfer mechanisms for specific nutrients. These deficiencies are intimately bound up with genetics, they are in fact inheritable diseases. Some humans have a rare defect of cation transport into their red blood cells while others have intestines that cannot transfer glucose against a concentration gradient. The defect, however, is highly specific, only the glucose mechanism seems to be affected, for such cells can still concentrate amino acids. Clearly it will be worth while to focus our attention on any differences between the chemical structure of these abnormal and normal cell membranes. We might well be able to isolate specific molecules in the normal that do not appear in the abnormal. Who knows, these may well be the elusive carrier structures for glucose.

### A final attempt at invention

The previous paragraphs have been three attempts to invent the future by picking research areas that show promise at this present time. This line of thought while useful is only too obvious. What we want is a way to stretch the mind even further. Can we break away from the massivity of the intellectual, educational and factual backgrounds that surround us and make a clear leap ahead, untramelled by present-day thinking? The answer, unfortunately, is probably in the negative for most of us, but one small ray of hope lies in inventing a university examination paper on membranes for the year 2000. What would the ten questions on such an examination paper contain? My guess is something like this:

# University of Newtown

## Faculty of Life Sciences

Mid-course examination for Batchelors Degree in Membranology (B.Memb.)

### Biological Membranes (Membranology – Descriptive Paper I)
July 2000, 9.30–12.30 a.m.

(STUDENTS MUST ANSWER AT LEAST THREE BUT NOT MORE THAN FOUR
QUESTIONS)

1. Show how Eiseman's method of grafting on a foreign cell membrane to a stripped cell (one produced by removing its own membrane) has been an enormously useful technique for demonstrating membrane control of cellular metabolism.
2. Discuss how the use of "living cell" electron microscopy has enabled us to understand the mechanics of the flowing membrane in detail.
3. Explain the membrane mechanisms by which individual cells recognize one of their own kind and describe how we can block this recognition process to our advantage for tissue transplantation surgery.
4. Describe the advances that have taken place in the methods of synthesizing chemical tags for specific structural groups in membranes.
5. Discuss Zemmerfield's artificial, self-reproducing, mobile membrane model and show how it differs from normal cell membranes.
6. Explain in detail the Narengenfeldt–Kohn–Hudson pathways by which plasma membrane sub-units are synthesized and describe how they are inserted and removed from the cell membrane proper.
7. Discuss how the isolation of the membrane carrier for glucose was carried out using mucosal cells of the small intestine as the test tissue.
8. List and briefly discuss the basic mechanisms by which hormones control cell function through changes in membrane permeability. How many secondary intracellular messengers are known? Give the evidence for their actions.
9. Briefly review our new concept of membrane action of anaesthetics and show how the early theories of anaesthesia were erroneous.
10. Discuss and detail the chemical reactions that link the hydrolysis of ATP with the membrane structures that actually carry out osmotic work (i.e. active transfer of solutes across membranes).

Perhaps my Newtown University examination paper in membranology is a disappointment, the reader seeing better questions than those set. If this is so all is well and good. For in a way, the ultimate pleasure for an author (apart from good royalties) is that his book has stimulated the reader to have thoughts where no thoughts were before! If you can invent better questions than those of Newtown's University you might well be the one who ends up finding their answers. You might not know exactly where you are going, but you are on your way!

# Bibliography

The references listed below are by no means a full list of all the sources consulted during preparation of this book. They represent, however, either the more important primary references on membranes, or books that contain an extensive survey of the field. Those articles and books that contain useful summaries are marked with an asterisk.

BENNETT, H. S. (1956). "The concepts of membrane flow and membrane vesiculation as mechanisms for active transport and ion pumping". *Journal of Biophysics and Biochemical Cytology*, **2** (Supplement), 99.

*BRACHET, J. (1961). "The Living Cell", *Scientific American* (September).

*BRACHET, J. and MIRSKEY, A. E. (1959–65), editors. *The Cell; Biochemistry, Physiology, Morphology*, Vols. 1–6, Academic Press, New York.

*BROWN, R. and DANIELLI, J. F. (1954), editors. "Active transport and secretion", *Symposia of the Society for Experimental Biology*, No. VIII, University Press, Cambridge.

*CHRISTENSEN, H. N. (1960). "Reactive Sites and Biological Transport", in *Advances in Protein Chemistry*, **16**, 239. Edited by Anfinsen, C. B., Anson, M. L., Bailey, K. and Edsall, J. T. Academic Press, New York.

COLLANDER, R. and BARLUND, H. (1933). "Permeabilitats-studien an Chara ceratophylla", *Acta Botanica Fennica*, **11**, 1.

CURRAN, P. F. and MACINTOSH, J. R. (1962). "A model system for biological water transport", *Nature* (London), **193**, 347.

DANIELLI, J. F. and DAVSON, H. (1935). "A Contribution to the Theory of the Permeability of Thin Films", *Journal of Cellular and Comparative Physiology*, **5**, 495.

*DAVSON, H. (1959). *A Textbook of General Physiology*. J. and A. Churchill Ltd., London.

*DAVSON, H. and DANIELLI, J. F. (1943). *The Permeability of Natural Membranes*. University Press, Cambridge.

*DIAMOND, J. M. and TORMEY, J. McD. (1966). "Studies on the Structural Basis of Water Transport across Epithelial Membranes", *Federation Proceedings of the Federation of American Societies for Experimental Biology*, **25**, 1458.

*DICK, D. A. T. (1966). *Cell Water*. Butterworths Molecular Biology and Medicine Series, London.

DIEDRICH, D. (1966). "Glucose transport carrier in dog kidney; its concentration and turnover number", *American Journal of Physiology*, **211**, 581.

*DOWNES, H. R. (1963). *The Chemistry of Living Cells*. Longmans, Green & Co. Ltd., London.

EDWARDS, J. G. (1925). "Formation of food-cups in amoeba induced by chemicals", *Biological Bulletin*, **48**, 236.

*ELBERS, P. F. (1964). "The Cell Membrane; Image and Interpretation", in *Recent Process in Surface Science*, **2**, 443. Edited by Danielli, J., Pankhurst, K. G. A. and Riddiford, A. C. Academic Press, New York.

*FINIAN, J. B. (1966). "The Molecular Organisation of Cell Membranes", in *Progress in Biophysics and Molecular Biology*, **16**, 145. Edited by Butler, J. A. V. and Huxley, H. E. Pergamon Press, Oxford.

*FOGG, G. E. (1965), editor. "The State and Movement of Water in Living Organisms", *Symposia of the Society for Experimental Biology*, No. XIX. University Press, Cambridge.

GOLDACRE, R. J. (1952). "The Folding and Unfolding of Protein Molecules as a Basis of Osmotic Work", in *International Review of Cytology*, **1**, 135. Edited by Bourne, G. H. and Danielli, J. F. Academic Press, New York.

GORTER, E. and GRENDEL, F. (1925). "On Bimolecular Layers of Lipoids on the Chromocytes of Blood", *Journal of Experimental Medicine*, **41**, 439.

GOSSELIN, R. E. (1967). "Kinetics of Pinocytosis", *Federation Proceedings of the Federation of American Societies for Experimental Biology*, **26**, 987.

*GREEN, D. E. and GOLDBERGER, R. F. (1967). *Molecular Insights into the Living Process*. Academic Press, New York.

*GROSS, L. (1967). "Active Membranes for Active Transport", *Journal of Theoretical Biology*, **15**, 298.

HOLTER, H. (1959). "Pinocytosis", in *International Review of Cytology*, **8**, 481. Edited by Bourne, G. H. and Danielli, J. F. Academic Press, New York.

JOHNSTON, P. V. and ROOTS, B. I. (1968). "Changing Views of Biological Membranes", *Medical and Biological Illustration*, **18**, 110.

KATZ, A. I. and EPSTEIN, F. H. (1967). "The Physiological Role of Sodium–Potassium Activated Adenosine Triphosphatase in Active Transport of Cations across Biological Membranes", *Israel Journal of Medical Sciences*, **3**, 155.

KOEFOED-JOHNSEN, V. and USSING, H. H. (1958). "The Nature of the Frog Skin Potential", *Acta Physiologica Scandinavica*, **42**, 298.

*KORN, E. D. (1966). "Structure of Biological Membranes", *Science*, **153**, 1491.

KORN, E. D. (1968). "Structure and Function of the Plasma Membrane", *Journal of General Physiology*, **52**, 257s.

LOWE, A. G. (1968). "Enzyme Mechanism for the Active Transport of Sodium and Potassium Ions in Animal Cells", *Nature* (London), **219** 934.

*MADDY, A. H. (1966). "The Chemical Organization of the Plasma Membrane of Animal Cells", in *International Review of Cytology*, **20**, 1. Edited by Bourne, G. and Danielli, J. F. Academic Press, New York.

*NORTHCOTE, D. H. (1968), editor. "Structure and Function of Membranes", *British Medical Bulletin*, **24**, 99.

*O'BRIEN, J. S. (1967). "Cell Membranes – Composition: Structure: Function", *Journal of Theoretical Biology*, **15**, 307.

PROSSER, C. L. (1952), editor. *Comparative Animal Physiology*. W. and B. Saunders Co., Philadelphia.

ROBERTSON, J. D. (1962). "The Membrane of the Living Cell", *Scientific American* (April).

ROBERTSON, J. D. (1964). "Unit Membranes: A Review with Recent New Studies of Experimental Alterations and a New Sub-unit Structure in Synaptic Membranes in *Cellular Membranes in Development*, (22nd *Symposium of the Society for the Study of Development and Growth* 1963). Edited by Locke, M. Academic Press, New York.

*ROBERTSON, J. D. (1966). "Unit Membrane and the Danielli–Davson Model", in *Intracellular Transport*, (*Symposia for the International Society of Cell Biology*), **5**, 1. Edited by Warren, K. B. Academic Press, New York.

RUSTAD, R. C. (1964). "The Physiology of Pinocytosis", in *Recent Progress in Surface Science*, **2**, 353. Edited by Danielli, J. F., Pankhurst, K. G. A. and Riddiford, A. C. Academic Press, New York.

SBARRA, A. J., SHIRLEY, W. and BARDAWIL, W. A. (1962). "'Piggyback' phagocytosis", *Nature* (London), **194**, 255.

SCHMIDT, W. J. (1936). "Doppelbrechung und Feinban der Markscheide der Nervenfasern", *Zeitschrift fur Zellforschung und mikroscopische Anatomie*, **23**, 657.

*SCHOFFENIELS, E. (1967). *Cellular Aspects of Membrane Permeability*. Pergamon Press, Oxford.

SJÖSTRAND, F. S. (1967). "Molecular Structure and Function of Cellular Membranes", in *Protides of the Biological Fluids* (Section A), **15**, 15. Edited by Peeters, H. Elsevier Publishing Company, Amsterdam.

*STEIN, W. D. (1967). *The Movement of Molecules across Cell Membranes*. Academic Press, New York.

STEIN, W. D. and DANIELLI, J. F. (1956). "Structure and Function in

Red Cell Permeability", *Discussions of the Faraday Society*, **21**, 238.

STIRLING, C. E. (1967). "High Resolution Radioautography of Phlorizin-$^3$H in Rings of Hamster Intestine", *Journal of Cellular Biology*, **35**, 605.

*TAYLOR, G. R. (1967). *The Science of Life – A Pictorial History of Biology*. Panther Books Ltd.

THOMPSON, T. E. (1964). "The Properties of Bimolecular Phospholipid Membranes", in Cellular Membranes in Development (*22nd Symposium of the Society for the Study of Development and Growth* 1963). Edited by Locke, M. Academic Press, New York.

USSING, H. H. and ZERAHN, K. (1951). "Active Transport of Sodium as the Source of Electric Current in the Short-Circuited Isolated Frog Skin", *Acta Physiologica Scandinavica*, **23**, 110.

*WHITTAM, R. (1964). "Transport and Diffusion in Red Blood Cells". Monographs of the Physiological Society No. 13. Edward Arnold Ltd., London.

WIDDAS, W. F. (1963). "Permeability", in *Recent Advances in Physiology*, 8th edition, 1. Edited by Creese, R. J. and A. Churchill Ltd., London.

WILSON, T. H. (1962). *Intestinal Absorption*. W. B. Saunders Co., Philadelphia.

# Index

# Index